DELIVER!

DELIVER!

The Mindset of High-Performing Project Managers

Brigadier General Greg Gutterman
United States Air Force, Retired

A Gutt Check Series Book
Greg Gutterman Group Publication

Greg Gutterman Group, LLC
P.O Box 94
Alpha, OH 45301

LIBRARY OF CONGRESS CATALOG-IN-PRINTING DATA
Gutterman, Gregory 1966 -
    Deliver! The Mindset of High-Performing Project Managers by Greg Gutterman. -- 1st ed.

ISBN: 979-8-9869443-3-3 (paperback) | 979-8-9869443-4-0 (ebook)

1. Production Operation and Management. 2. Industrial Production Engineering. 3. Project Management Technical. 4. Engineering Management. 5. Military Technology. 6. Project Management. 7. Project Management Business. 8. Conventional Weapons and Warfare History. 9. Engineering Design. 10. Manufacturing. 11. Adult and Continuing Education.

Cover design by The Next Wave, LLC
Copy Editor Larkin Vonalt
First Edition. Printed in the United States of America

The views expressed in this publication are those of the author and do not necessarily reflect the official policy or position of the Department of Defense or the U.S. government. The public release clearance of this publication by the Department of Defense does not imply Department of Defense endorsement or factual accuracy of the material.

For our nation's warriors,
lost heroes, and their selfless families.

Special Note from the Author

A portion of the proceeds from this book will be donated to the Wounded Warrior Project®. Their mission is to honor and empower wounded warriors who incurred physical or mental injury, illness, or wounds co-incident with military service on or after September 11, 2001. For more information, visit www.woundedwarriorproject.org.

Table of Contents

Preface

Building 28 at Wright-Patterson Air Force Base in Ohio was home to some of the most highly classified weapon development projects in the United States Air Force. It was there I first learned about Big Safari. The top-secret project management office spent forty years in the shadows rapidly developing some of the Cold War's most technologically advanced intelligence, surveillance, and reconnaissance weapon systems. In the early 1990s, the Air Force decided to shine a light on Big Safari and absorbed the mavericks into a larger organization. Bill Grimes, the long-time Big Safari Program Director, was not happy. "These bureaucratic pukes are trying to destroy us" he'd openly and unapologetically say to anyone listening.

I had the pleasure of witnessing Grimes' cold receptions numerous times hitchhiking to meetings in support of my then-boss. Sitting near the back of the room unnoticed, I'd hear Grimes whisper "ridiculous" under his breath time and again. The reason? A non-Big Safari project manager was on stage selling their project to leadership, instead of providing evidence the project was being proactively managed. Grimes' verbal and non-verbal communication was unintentional instructions for anyone paying attention. I paid close attention to Big Safari's techniques. Over time the way to bend customers' requirements to fit a technological environment, plow bureaucratic roadblocks, unify key stakeholders, and lead project teams became crystal clear.

A decade after my time in building twenty-eight, I was fostering the same project management principles on a satellite-guided bomb program. The record-breaking project was delivered with Big Safari speed and precision outside the Pentagon's streamlined top-secret processes. The same recipe was used to deliver an enterprise software system, resolve the perceived hypoxia concerns on the F-22 Raptor stealth fighter, and change the culture of the Air Force's Foreign Military Sales enterprise. The formula worked at every stop.

After more than three decades leading successful product development teams, it is clear accelerating delivery of projects and products is not a process; it's a mindset. A mindset comprised of the ten core principles contained in this intentionally concise book. If you want to learn how to avoid the bureaucratic quicksand and charge ahead of your competitors like the high-performing project managers do, read on!

Introduction: Don't Let *That* Happen

My project management compass developed quickly. My teacher was a five-foot-three-inch tall thirty-one-year-old brunette. She had the audacity to apply for an open position at a military veterans-only bar.

"Listen, I don't care if she's a woman," the conversation would start. "The way I see it is, if she can't be drafted and sent into combat like we were, she should not be serving us drinks!" The majority opinion was meant to mask the truth. The boys were worried their tipsy chatter would be broadcast to their wives by the female spy in the bar.

Mom needed the job, so she dug in. Her persistence and some behind-the-scenes diplomacy by my dad paid off. In the late 1970s, Mom was hired as the first-ever female bartender at American Legion Post 98 in Saint Paul Park, Minnesota. There has been a female tending the bar ever since. Mom shattered their glass ceiling.

Mom was a quick-witted sensation at the pub and before my brothers and I knew it, she passed the "bar." She was both a taproom lawyer and psychologist. I learned about her informal education during one of my infrequent phone calls home after joining the military as a research and development engineer.

Mom answered the phone and got right to the point. "Gregory, did you hear about the five-hundred-dollar hammer the Air Force bought? I think a hammer is a dollar at the hardware store around the corner. Should I mail you one?" she said with soaking sarcasm.

"Yeah, Ma, I read about that," I replied.

"Well, how could you let that happen?" she asked.

"Mom, I didn't ..." I stammered, but she cut me off.

"Don't engineers design and develop weapons and equipment for the Air Force?" Mom asked. I was cornered and the final uppercut was about to be thrown.

"Gregory, you're an Air Force Officer, a leader, and five hundred dollars of taxpayer money was wasted on that hammer. I'll ask you again: how could you let *that* happen in *your* Air Force?" With that Mom's point was made and her case closed.

It was my Air Force and my new profession being hammered by Mom. I soon realized Mom was just telling me what the military veterans in the bar and most of our nation's taxpayers were thinking: purchasing a five-hundred-dollar hammer was wasteful government spending. Whether it was justified or not they needed a scapegoat. I was the Air Force in their eyes and the only one they could hold even remotely accountable for the wasteful government spending.[1]

Recognizing the nation's taxpayers had this perspective drove me to adopt "I won't let *that* happen" as my project management moral compass. This helped me become more inquisitive, pay closer attention to the details, and take on more personal accountability. In time I learned how to question foolish requirements, set and meet commitments, and navigate around unhelpful bureaucracy. Mom's simple guiding principle focused me on a higher-level obligation to our nation's taxpayers and warfighters. She was right, it was *my* responsibility to develop and deliver war-winning capabilities without waste or excuses.

Twenty years later I landed my dream job: System Program Director of the United States Air Force's sixty-billion-dollar F-22 Raptor jet fighter. Raptor is the world's first fifth-generation stealth fighter and the most technologically advanced airplane on the planet. The new job was equally gratifying and horrifying.

A year earlier an F-22 pilot lost his life in a terrible aircraft accident. The crash captured national attention and with-it numerous Air Force and Department of Defense-level investigations. Despite thousands of hours of technical analysis, the cause of the accident remained unknown. The Air Force responded by limiting F-22 flights to non-combat training missions only: Raptor's powerful wings were clipped.

The direction from above was unambiguous: find the technical problems, fix them, and quickly get the F-22 Fleet back to full combat capability. There was an unspoken and equally daunting task: restore F-22 pilots' trust in their jet and recapture their confidence in the program's technical team too.

Before the "new guy" program and team introductions were complete, the commotion spilled over. Two F-22 pilots appeared on a nationally televised news program to openly air their distrust in the jet's life-support systems. The pilots were convinced, as were others, that oxygen deprivation or hypoxia caused the accident. Once a person locks in a belief, it is hard to convince them otherwise even with compelling evidence.

I read all available reports on the F-22 in preparation for the job. Every report had a different conclusion about the possible cause of the accident and different recommendations on aircraft system modifications as a result. The United States Air Force's Accident Investigation Board Report, though, had me once again recalling Mom's early career guidance.

The F-22 accident report indicated the onboard oxygen generation system failed before the crash. This system captures air from outside the aircraft and transforms it into breathable oxygen for the pilots, just like commercial airplanes do for their passengers. However, unlike commercial airplanes where breathing air is vented into the passenger main cabin, fighter pilots breathe oxygen from a scuba-diver type mask covering their nose and mouth. When the onboard oxygen system stopped sending breathing air to the mask, the pilot entered an in-flight emergency nicknamed "sucking rubber."

What happened next inside the cockpit will remain a mystery forever. The accident report indicates the F-22 pilot did not manually activate the backup oxygen tank inside the cockpit or declare an in-flight emergency. Both steps are required per F-22 in-flight emergency protocols when the onboard oxygen system fails; a process the pilot was well-trained to follow.

Reading the accident investigation board's report reddened my face. It was my Air Force acquisition and engineering community that decided to install a manually activated backup breathing system on the world's most sophisticated fighter aircraft. A jet that would put almost unbearable gravitational forces on its pilots. In a sucking rubber scenario, we expected a pilot to manually turn on a backup breathing apparatus.

I wrote "How could we let that happen? We can and must do better!" on the front page of the accident investigation report. I placed it on the corner of my desk as a daily reminder of our nation's lost pilot and his family, and why proper product development is so critically important to those we are honored to serve.

A priority for the F-22 product development team immediately became installing an automatic backup oxygen system on the F-22 Raptor. A small, experienced, and determined team of A-players was assembled and my marching order, or project intent, was unambiguous.

> *"Deliver an automatic backup oxygen system on the F-22 within twelve months. I know it's aggressive, but I have your back. I will plow the bureaucratic roadblocks that get in your way. I know you can do it, and we owe it to the pilots and their families to get it done quickly."*

Within a week the automatic backup oxygen system team came into my office to tell me twelve months to initial delivery of the system was no longer possible. They hit their first major bureaucratic obstacle, the Raptor's requirements team at Langley Air Force Base, Virginia.

The Raptor's maintenance representatives wanted to add a new requirement to the list; the ability to refill a depleted backup oxygen tank without removing it from an F-22 cockpit. The goal, they cited, was to refill depleted oxygen tanks on the ground while the jet was still running. The new requirement was intended to decrease aircraft downtime between missions, especially in wartime situations.

"How do we refill an empty backup oxygen tank today?" I asked the team. The answer was reminiscent of refilling a gas grille's propane tank at the local hardware store. The empty tank is taken off the jet and swapped out for a pre-filled oxygen tank. A simple and fast procedure. The empty oxygen tanks are then taken to an environmentally friendly room well away from any potential airborne contaminants on the flight line and refilled with oxygen. This back-shop process ensures the pilots' breathing air is environmentally pure and safe.

The flight-line oxygen tank refilling requirement made little sense, and the team needed my top cover.

> *"Tell Langley "No." We are not going to add this requirement. We will refill the oxygen tanks exactly as we are doing it today. We are going to field this capability in twelve months. We owe it to the pilots and their families."*

> *"If Langley feels I am overstepping my authority, great. They can overrule me in writing. But it must be signed by a General Officer. In the meantime, press on!"*

I had drawn a line in the sand and the blow-back started immediately. My response to the "Who do you think you are?" electronic mail and phone calls was the same every time: "I understand, please put that in writing and have it signed by a general officer."

My prior staff assignments taught me how hard it was to get a military general officer (a corporate senior executive) to sign anything, even a birthday card. The Raptor requirements team would have to get on their general officer's calendar, explain the situation, and then staff a letter through numerous layers of review and editing. The letter would be sent from Langley Air Force Base in Virginia to my Commander in Ohio. The two military general officers would then have a phone conversation about the "belligerent colonel." After which my general officer would talk to me in person, hand me the Langley letter, and say, "Do it." My estimate was it would take four months for this process to play out. I was wrong, it took six.

When my superior gave me his direction to add the flightline refilling requirement, I immediately obliged.

"Yes Sir! We will add the capability in Phase II. Phase I is in final testing and the fleet retrofit is starting within sixty days."

My commanding general officer agreed to delay the scope expanding requirement, relayed the strategy to the Langley military general officer, and the team started fielding the F-22 Automatic Backup Oxygen System within twelve months as promised.

The requirements battle was lost, but the pilots won the *needed* capability. Most importantly, a pilot would never be forced to suck rubber, or put breathing ahead of flying, again.

The F-22 pilots applauded the addition of an automatic backup oxygen system on their fighter. But an equally important outcome resulted from meeting our delivery commitment: we recaptured their trust. This is an example of successful product development by any measure.

Rapidly moving the F-22 program[2] to this point was accomplished by fighting for and demanding achievable and realistic requirements, setting a program intent, avoiding scope creep, and meeting commitments. These are a few of the ten project management principles comprising the mindset of high-performing project managers. Project managers who avoid the bureaucratic quicksand and rapidly deliver value-added products to their customers. Although the examples used throughout this purposefully short book relate to developing military systems, the principles apply to all project management professionals, irrespective of industry or specialty.

There is fierce competition in every industry including nation states. The future winners will be the companies and countries able to introduce new and innovative products or weapons of war quickly. High-performing project managers know how to deliver value-added products rapidly, because they know it is the difference between winning and losing.

"Victory smiles upon those who anticipate the changes in the character of war, not upon those who wait to adapt themselves after the changes occur."
- Italian Air Marshall Douhet, 1928

Chapter Endnotes

[1] The cost of the specialized hammer included non-recurring engineering and corporate overhead charges. But as the saying goes, one person's perception is another's reality.

[2] The terms "program" and "project" are used interchangeably throughout this book. A program is comprised of two or more subordinate projects. Likewise, the title "program manager" identifies a supervisor overseeing two or more related projects (and project managers). A diverse number of projects are led by portfolio managers.

First: Learn to Navigate

Identify and avoid the
bureaucratic roadblocks

The perception that developing the machines of war takes too long and costs too much is unfortunately true. This fact is one of the reasons Department of Defense political appointees continually attempt to make their mark by implementing some form of acquisition process reform.[3] The two most recent transformation efforts were dubbed "Lightning Bolts" and later "Better Buying Power" initiatives.

In fairness, some of these efforts did result in short-term cost or schedule improvements for a few select programs. This was primarily due to the fact the bureaucracy supports dramatic tailoring of the process for any program under the watchful eyes of a Pentagon senior executive or political appointee. But as soon as the spotlight is removed the bureaucratic "kiss-the-ring" process returns.

This fact is well documented. A great read on the topic is *Defense Acquisition Reform 1960-2009, An Elusive Goal*. Ironically the historical review was expected to cover a longer time horizon, but the study fell prey to a flaw in the system. Personnel turnover led to priority changes, priority changes to loss of funding, and loss of funding truncated the project.

There is one common denominator resulting from every reform effort undertaken highlighted in the acquisition reform book: additional oversight. Oversight has a rightful purpose, but the ever-expanding amount of oversight of the acquisition (or product development) process has clearly led to increased costs and elongated schedules.[4]

The average large-dollar Major Defense Acquisition Program has historically consumed fifty percent more funding and taken a full two years longer than originally projected.[5] This finding is from the year 2020, but the data has been the same for decades. Government data clearly indicates major defense acquisition program budgets and schedules are expanding more than ever before.

One recent government report indicated six of seven major weapon system development programs requesting additional funding to cover budget overruns in the year 2021, needed even more money in 2022. For schedule delays, the same situation exists: nine of the nine major programs that reported schedule delays in the year 2021, reported delays again in 2022.[6] The cost and schedule problems in the defense acquisition process persist despite numerous reforms and added oversight.

This pervasive re-baselining of major program cost and schedule targets is the result of good intentions gone wrong. Congress, justifiably frustrated by perpetual cost and schedule issues on major development programs, enacted enhanced oversight. The statute sets limits on acceptable program cost and schedule growth and automatically triggers a significant inquiry once the limits are breached.[7]

Tiered oversight, however, has made the cost and schedule situations worse. Pragmatic program managers now routinely add more time and budget to their product development estimates at inception. By doing so they hope to avoid a Congressional program breach in the first place. This unintended consequence is known as the "zero-risk" program plan, a practice now deeply ingrained into the culture of the Defense Acquisition System. The zero-risk program plan undergoes significant scrutiny during the bureaucratic review process. Again, the additional oversight further inflates an already bloated project plan. The finalized cost and schedule targets then become self-fulfilling prophecies. Navigating this bureaucratic quicksand begins by understanding the three reasons it exists in the first place: the veto, unnecessary compromise, and the statutory division of power.

Veto power is a form of informal control exerted by well-intentioned bureaucrats whose coordination is required to secure program resources and approvals. This problem normally appears during the document review process in the Pentagon. Let's examine one of the numerous documents required to obtain permission to proceed with a product development program: the Acquisition Strategy Document. After the strategy document is moved into the Pentagon, the first step is to capture the specific military departments' (e.g., Department of the Air Force) approval.

For the Department of the Air Force, the product development decision maker is the Assistant Secretary of the Air Force for Acquisition, Technology, and Logistics. Moving the strategy document to this level, however, requires coordination from ten subordinate offices first. This is where the Pentagon's project document staffing process gets cumbersome. Every one of the Assistant Secretary's subordinate offices requires three levels of review: entry, intermediate, and senior. Just to be clear, this is ten offices with three layers of review each.

During the entry-level review process, junior staffers review the document and more often than not, request edits. Editing the document, however, requires re-coordination with the nine other subordinate offices just in case they disagree with the update. Refusing to accommodate the reviewers' edit results in an all-stop documentation veto. More on that later.

Once this back and forth is completed at the entry level, the process is repeated at the intermediate level with the same ten offices. After the intermediate-level documentation reviews, tweaks, and modifications are completed, another review is kicked off. Same ten offices, but this time at the senior level of review. Senior-level coordination finally allows the project document to make it to the Assistant Secretary; the service-level decision maker.

The Assistant Secretary's signature authorizes the document to move up to the ultimate Department of Defense decision-maker; the Undersecretary of Defense for Acquisition and Sustainment. Once again, the three-tiered review and approval process is repeated, this time with the numerous subordinate offices reporting to the Undersecretary of Defense.

This coordination, review, and approval process is repeated for every one of the tens of documents required to capture the authority to proceed with a weapon or support system design, development, and production effort. This document review process takes months, not days.

As cumbersome as the process may appear, the process is not the problem. The problem is every person at every level in the bureaucracy believes they have the power to veto your work. Too many program managers unknowingly permit this to occur, because reviewers do not have the power to veto anything.

Here's an example. An Air Force bomb program's Acquisition Strategy Document finally made it through Air Force staffing and moved into the Office of the Secretary of Defense for approval. The bomb team thought they were home free, but they were not. During the intermediate-level review, the Undersecretary of Defense's test and evaluation representative requested a significant document edit. Coordination on the document would be withheld until the requested edits were inserted into the document's test section. By withholding a signature, the Undersecretary of Defense's test representative issued a process veto bringing the entire staffing process to a complete stop.

The requested edit to the test section was considered a major update to the bomb's acquisition strategy document. An update requiring full re-coordination with the ten-subordinate offices a reporting layer below the Undersecretary of Defense, back at the Department of the Air Force level. A significant late-process edit like this delays the approvals required to move the program forward into the design and/or development phase. Even worse, Pentagon financial accountants would view the staffing delay as an indication the bomb program would not need the set-aside budget until a later date too. Before you know it, the bomb's budget is reallocated to a lower-priority but ready-to-execute program, thereby further delaying the project. Yes, a simple document staffing issue virtually guarantees the program will be delayed for at least a year due to funding reallocations. The bottom line is a process veto delays schedules, increases costs, and even puts programs at risk of termination via funding cuts.

Surrendering to the demands of the bureaucracy is the norm. A traditional program manager receives a critical comment and immediately agrees to update the document. This starts the time-consuming, schedule delaying, and cost-increasing re-coordination process. Unfortunately, this happens with little to no debate.

Here's a dirty little secret. The coordinating officials in the Pentagon do not have the power to issue a veto, say no, or hold up program progress. The 'yes' or 'no' power rests with a select few designated as Decision Authorities.[8] These high-placed executives are the only ones with veto power. Unfortunately, the average program manager doesn't recognize this important fact, and their program suffers elongated schedules and increased costs as a result.

The Air Force bomb program team, however, knew where veto power rested. Instead of caving to the no-value added test section update, they told the

Undersecretary of Defense's test representative "No." Then the program manager led an unemotional fact-based conversation about the merits of their response.

The bomb program manager reminded the Undersecretary of Defense's test representative the content requested was in a separate, but related document; the Test and Evaluation Master Plan. Duplicating test-related information in two or more documents creates inconsistencies and confusion. When this occurs, the test community will not use the Acquisition Strategy to resolve the conflict. Instead, everyone will point to the authoritative test planning document: the Test and Evaluation Master Plan. This was a point the program team and Undersecretary of Defense's test representative agreed upon. This honest and thoughtful dialogue between the Air Force's bomb program manager and the Undersecretary's test representative resolved the veto. The request to update the test section was scrapped saving months of re-coordination time and potentially the program.

The average program manager's willing acceptance of a process veto leads to the second major problem in the Defense Acquisition System: unnecessary compromise. As of this writing, nearly all of the documents required for authority to proceed in the product development process make it to the ultimate decision maker without a single contrarian view. That is, strategies and plans for multi-million or multi-billion-dollar product developments are being watered down in an attempt to satisfy everyone on the Pentagon staff; everyone.

This problem is deeply ingrained because the military's project management community requires conflict to be worked out at a lower level to "protect the boss." When this occurs, the true decision-maker is denied varying points of view and essential information. Unnecessary compromise unintentionally encourages organizational groupthink, instead of healthy debate and data-driven decision-making.

Here's an example of this particular bureaucratic problem. An Air Force enterprise software system was nearing a major Pentagon decision point known as Milestone B, the program of record initiation. This milestone is a big deal because at program initiation a Congressional checking account is opened up for a program.

After obtaining the Air Force's approval, the strategy document entered the Undersecretary of Defense for Acquisition and Sustainment's office for review and final approval. During subordinate office review, however, the Undersecretary's engineering representative issued a process veto. They refused to coordinate on the acquisition strategy document until the program manager submitted a separate and independent software engineering-focused document. This is bureaucratic hostage-holding.

The software system's program manager notified the Air Force about the demand for a new document and requested a twelve-month delay to the program initiation decision. The year-long delay was required, the program manager said, to create an expected six-hundred-page software engineering-focused report. The Air Force's lead staffing agent in the Pentagon reminded the program manager a single-page document took months to coordinate in the building, a six-hundred-page technically-focused document would take years. The significant delay would trigger Pentagon accountants to reallocate the budget of the program thereby effectively canceling the product development altogether.

The Undersecretary of Defense's engineering representative's request for a separate document had nothing to do with the software system's acquisition strategy; the request added zero value. When the process becomes so blatantly anti-product, the bureaucratic beast must be told "No." The Air Force refused the document blackmail, and eventually, the Undersecretary of Defense's engineering representative grudgingly coordinated on the strategy document. The software system received approval to proceed to program initiation a few months later as planned.

The culture of compromise is driven in large part by the third major problem within the Defense Acquisition System: statutory call-outs. The statutory call-out problem occurs when Congress grants a subordinate office inside the Defense Department as much formal power as their superior. This is done by embedding language into an annual National Defense Authorization Act. Several of the Undersecretary of Defense for Acquisition and Sustainment's subordinates wield this form of statutory power.

For example, Congress has directed the Undersecretary of Defense's Director of Operational Test and Evaluation to review and approve every major defense acquisition program's Test and Evaluation

Master Plan. The Director for Defense Procurement and Acquisition Policy is required to approve the contracting approach. The Director of Cost Assessment and Program Evaluation approves the program's cost estimate. And so on.

Congressional statutory call-outs provide the subordinate with as much legal power as the superior. The teenager can overrule the parent, and they do. Here's an example. The F-22 program was seeking approval for a classified project in the Pentagon. During a meeting with the Undersecretary of Defense for Acquisition and Sustainment an issue was raised concerning the project's cost estimate. The Undersecretary ultimately agreed with the Air Force and directed the staff to approve the Air Force cost estimate for the project.

Five months later the Undersecretary of Defense's decision was still not implemented and the F-22 project was stuck in limbo. The Undersecretary's Director for Cost Assessment and Program Evaluation had the statutory power to decide on the official cost estimate, and they quietly disagreed with their superior. Unbeknownst to the Undersecretary of Defense for Acquisition and Sustainment, the Air Force was forced to allocate more budget to the project. Once the Air Force did, the Director of Cost Assessment and Program Evaluation signed the necessary document to move the program forward.

Withholding a signature or coordination is the major implication of the statutory power call-outs. An Undersecretary of Defense subordinate office can place a "legal" demand on the program manager, and until they comply, the signature is withheld. This puts the program manager in a quandary: compromise with the demand to move the program forward, or say no to the demand and suffer an indefinite product development delay. When statutory power is used in a negative manner like this, nobody wins.

The three major problems within the Defense Department's product development process can be boiled down into a single sentence.

The statutory division of power emboldens Pentagon staffers at all levels to issue vetoes until the program manager caves to the demand or a compromise is reached.

This unintended consequence of additional oversight drives a full year or more into product development schedules for each major milestone. This is necessary to accommodate the three major Defense Acquisition System problems. Equally, the expanding timelines bring to life Parkinson's Law: work expands to fill the time (and budget) allotted.

Despite the problems highlighted in the stories above, never forget the United States enjoys the best-equipped and most powerful fighting force on the planet. This is because there are thousands and thousands of incredible people who work tirelessly inside the bureaucracy to equip our nation's warriors with war-winning weapon systems. These mavericks recognize the litmus test for accepting a veto and a compromise is to ask: is it in the best interest of our nation's warfighters and taxpayers? A veto and compromise adding value (e.g., personnel safety improvement) is necessary and must be adopted. A no-value-added veto, however, drives an unnecessary compromise and must be challenged. Honest and thoughtful debate on the merits of any documentation change request enables better decision-making and speeds product to the field. This is the mindset of high-performing project managers. A "debate is encouraged" mindset that should become the standard.

Politics, policy, and process exist in every company and every industry. Recognizing the real problems in the process emboldens the program team to better navigate roadblocks and avoid slipping into the bureaucratic quicksand. This all starts by focusing everyone involved on the one reason the process exists in the first place: to rapidly deliver capabilities. Every question concerning product development should be centered on one question: does the change enable rapid, affordable, and safe product delivery? If the answer is no, everyone involved should easily reject the change and move on. High-performing project managers keep this thought process simple: never accept the "No" of a veto from someone in the enterprise who is not authorized to tell you 'Yes.' Amen!

Recognizing the bureaucratic process problems in your company is the first step on your path to high performance. The remainder of this book contains time-tested strategies for navigating around the bureaucratic roadblocks you will inevitably encounter as a product development professional.

Chapter Endnotes

---

[3] Over twenty-five major acquisition process over-hauls have been implemented over the last four decades. Regardless, the systemic problems remain. See *Defense Acquisition Reform: An Elusive Goal*. The full citation is located in the bibliography.
[4] *Defense Acquisition Reform 1960-2009, An Elusive Goal*. See page 147. The full citation is located in the bibliography.
[5] Government Accountability Office's *Defense Annual Acquisition Assessment*. June 3, 2020. The full citation is located in the bibliography.
[6] *GAO Report to Congressional Committees,* Weapon Systems Annual Assessment. Pages 27 to 28. The full citation located in the bibliography.
[7] This is known as a Nunn-McCurdy Breach.
[8] The three Pentagon product development decision makers are the Milestone Decision Authority, Service Acquisition Executive, or the Component Acquisition Executive. The Pentagon's interest level and project's total cost are used to determine which of the three will serve as the decision authority.

Second: Define Success

Focus on product ahead of process

A few hand-picked military colonels and their civilian equivalents have a rare honor. They get to lead the teams responsible for developing million- or billion-dollar major defense acquisition systems. This impressive group of United States Army, Navy, Marine, Space Force, and Air Force program managers also have the opportunity to attend a mandatory three-week-long preparatory course in Washington, D.C.

Several years ago, the final lesson of the three-week-long course featured a guest lecturer. The well-regarded retired military general officer opened the session by asking: "What's the definition of success for a program manager and their team?"

Several brave souls attempted to answer the open-ended question, but were rebuffed by the instructor's "Wrong!" The resulting confused silence was eventually broken by the guest instructor, who enlightened the room by answering his own question.

> *"The definition of success for a program manager and team is to make sure their program does not get canceled."*

The guest lecturer's definition of success is not documented, but it is deeply embedded into the culture of the Department of Defense's product development corps. This is a significant issue.

Getting permission to spend corporate funding on a program is a big deal in any industry. When the military promotes an activity to "Program of Record" status, it is a big deal too. A Program of Record decision (also known as program initiation) graduates a design effort into a development effort. At that point, Congress commits to appropriating taxpayer treasure to fund the weapon or support systems entire product development lifecycle. Program initiation is a big, *big* deal.

The Program of Record's Congressional checking account is used to hire a primary contractor or company to undertake the product development effort. The lead company uses some of the windfall to buy the necessary product components (e.g., canopy, radio) from subordinate vendors or subcontractors. The subcontractors do the same. When this hiring ripple is complete, one Program of Record creates thousands of American jobs.

Jobs not just anywhere by the way. The new workload is spread out to as many congressional districts as possible. Why? Because elected members of Congress will not turn off a funding spigot to a program if it means jobs will be lost back in their home districts. Even programs with significant cost overruns and/or schedule delays capture this near-immunity. Becoming a Program of Record increases the probability that a governmental product development effort will "not get canceled." Is this really success?

The "not canceled" culture motivates program managers to drive their program forward to Program of Record status entering product development well before a design has stabilized or technology has matured. Unstable product designs are bound to change, and when they do extra time and money is consumed. Immature technology requires unpredictable amounts of trial and error to mature, which also elongates schedules and breaks budgets. The fact is, unstable product designs and immature technology are the top two causes of project cost overruns and schedule delays. Consciously or subconsciously, the "not canceled" cancer has metastasized.

Here's an example. The F-22 Raptor jet engine became a program of record and entered product development in the year 1991. By 1998 the program was plagued by significant cost overruns and schedule delays. A frustrated Pentagon eventually pinpointed the powerplant's problem: immature jet engine technology. The immature jet engine technology suffered numerous manufacturing and test failures. Manufacturing issues and test failures required root-cause analysis, analysis led to engineering design changes, and engineering changes required re-testing to verify they worked. This test-fix-retest engineering change process was the primary reason the engine's product development budget and schedule ballooned.

A single engineering change is like a pebble thrown into a calm pond. The resultant ripples affect downstream efforts forcing them to pause and wait until the change is implemented and the ripples dissipate. There were a lot of engineering change pebbles being thrown into the F-22 propulsion system development pond, and the cost and schedule ripples became surfable waves.

A single engineering change during product development has an unfavorable effect on cost and schedule for one reason: people. Programs are staffed to accomplish pre-existing or planned workloads. They are not staffed to handle pre-existing *and* engineering-change-driven workloads at the same time. This personnel shortage puts the project manager in a conundrum. The project manager can re-task existing personnel to work the new engineering change-driven workload and delay the schedule. Or, the manager can add more people onto the program team to work the engineering change-driven work in parallel with the pre-existing workload. It's a catch-22: either way, costs increase and schedules delay.

The F-22 engine vendor chose to do the latter, they increased the size of their team. The new team members were tasked to work the change-driven workload in an attempt to maintain the original development schedule. That is, F-22 engine program leaders added to the program's personnel costs in a vain attempt to save the schedule. This was done for every change pebble, and eventually, the team doubled in size. The payroll costs doubled too resulting in a forty-million-dollar cost overrun. Immature technology-driven changes created tidal waves of cost and schedule problems.

The F-22 engine product development story highlights what happens when program success is defined as "not canceled." Program managers rush to Program of Record status and begin engineering and manufacturing development prematurely. When they do, they have set the stage for costly and time-consuming engineering changes.

Regrettably the "not canceled" culture runs deep. The C-130J is a cigar-shaped cargo aircraft originally fielded in the 1960s. In 2008 a C-130J aircraft modernization program was approaching a Pentagon program initiation decision. During an Air Force preparatory program briefing, a military four-star general asked a telling question: "Are you really expected to reduce the C-130s radar cross section?"

When you hear the term "stealth" it is about an object's radar cross-section. Small radar cross-sections are less visible to radar detection. Reducing the radar cross section for a cigar-shaped 1960s-era cargo aircraft was impossible, and everyone in the room knew it. Nonetheless, the unachievable requirement was officially listed as a necessary program capability.

The C-130J modernization program manager's response to the general officer's question was telling.

"No Sir, we are working with the warfighting customer now and expect to eliminate the requirement after program initiation."

Fortunately, prudence prevailed. The program manager agreed to some top-cover and the military four-star general immediately made a phone call to his operational counterpart. The ridiculous C-130 stealth requirement was scrapped before the phone call ended. The project then went on to become the Pentagon's top-performing project two short years later.

How did this foolish requirement see the light of day? Because the "not canceled" definition of success consciously or unconsciously continues to rule the day. The blind push to become a program of record before the design is stable and technology has matured must end.

An effective project management professional focuses on *delivering* value-added products to the customer. They demand achievable requirements by focusing on the capability *needed* to close a threat gap or exploit a market opportunity. They recognize the only true measures taxpayers or corporate leaders use to track the performance of a program are (1) did the program deliver *on or under budget*, and (2) did the program deliver *on time or early*. As a result, highly effective project leaders reject the "not-canceled" process oriented definition of success, and adopt a product focused "*to deliver*" alternative.

> *The definition of success for a project or program is to deliver the needed capability on time and on cost.*

The word *needed* is used instead of *required* in this definition of success for a reason. Project management mavericks eliminate valueless and/or unproduceable requirements at the start of a project. This important principle is described in the next chapter.

The Small Diameter Bomb program adopted the "to deliver" definition of program success in the early 2000s. The bomb was delivered on time and nearly four hundred million dollars below budget. Their recipe "to deliver" successfully was, first, achieve technical maturity and design stability *before* the program of record or initiation decision. Second, achieve design stability and technical maturity by leveraging competition for as long as possible. These factors combined to eliminate the epidemic of late-breaking, time-consuming, and costly engineering changes altogether. *Achieving design stability is the foundational building block of product development success,* which is why the theme will appear again in subsequent chapters.

Accelerating product development and delivery requires everyone in the enterprise to focus on the same definition of success or end state. That end state is to deliver the needed capability to the customer on time and on cost. The process is used as a means to this end. High-performing project managers create this type of focus, effort, and unity by continually placing product delivery at the forefront of all deliberations.

Third: Demand Achievable Requirements

Leverage existing technology, dodge inventing, and avoid scope creep

A classified camera program for a high-altitude reconnaissance and surveillance airplane received a rare honor: the program was terminated by Air Force leadership. Programs do not die instantaneously. Contractual agreements must be reconciled and outstanding bills must be paid. For the camera program, the contract close-out process took twelve months. Although the government may move deliberately to close out a terminated contract, the affected companies do not. The pink slips start flowing immediately. The laid-off employees receive no notice and upon termination are swiftly escorted out of the facility by security personnel. This intentionally unempathetic process has a purpose: to make sure emotionally distraught employees do not destroy property or steal classified information on their way out.

Terminated programs negatively impact three important stakeholder groups. First, the customer is denied a needed capability. Second, the project manager's resume' is professionally stained. Finally, numerous people instantly lose their livelihoods. How could the project manager let that happen?

The high-altitude aircraft-mounted camera was required to take panoramic pictures by rotating from side to side in ten seconds as the aircraft flew along its pre-planned flight path. Seemed simple enough, but it wasn't. During developmental testing the panoramic rotation caused the camera to vibrate rendering the images blurry and unusable.

The program engineers experimented with different gimbals, motors, and power supplies, but the pictures remained blurry. The only tweak that worked to dampen the vibrations was to rotate the camera five seconds slower. The slower rotation dampened the vibrations and clear pictures resulted. However, every attempt by engineering to change the side-to-side rotation requirement from ten seconds to fifteen seconds was unsuccessful. The camera was required to rotate in ten seconds, and "A requirement is a requirement" they were told.

A vibration-free ten-second rotation required an invention, an entirely new way of mitigating camera vibrations. Dampening springs proved non-viable, so the engineering team began working to invent a magnetically levitating camera mount. The magnetic connection would create a physical separation isolating vibrations to the mount but not transferring them to the camera. Brilliant. But it had never been done before and required an invention. The unexpected test-fix-retest trial-and-error process of inventing drove up program costs and delayed schedules. Eventually, the over-budget shock waves in the Pentagon were too much and the program was canceled.

A post-mortem analysis answered the "why ten-seconds" question. "We already have a camera that rotates in fifteen seconds," the customer said, "Ten seemed like an improvement." No operational reason. No value-added explanation. No additional photographic capability was provided. The ten-second rotation requirement was ill-defined, valueless, and unachievable. Is a requirement like this a good requirement?

Product development efforts suffering continual cost and schedule problems typically have an unachievable requirements problem. The hopeful quest to solve the unsolvable requires an unpredictable trial-and-error process, and the numerous design changes along the way increase costs and delay schedules. The reality is, projects are not staffed or funded to invent new technology. Effective project management professionals avoid this dilemma altogether. They demand achievable requirements to leverage existing technology thereby speeding products through the design and development process.

Demanding achievable requirements worked exceptionally well for the Small Diameter Bomb development in the early 2000s. Like many projects, the team started with an unachievable requirement. The warfighting customers wanted to put an F-15 Eagle jet fighter into a nosedive toward the ground, then at 10,000 feet above ground level, release a satellite-guided bomb. After releasing the bomb, and with only seconds to spare, the pilot would abruptly begin a vertical climb to avoid hitting the ground. The released bomb was then expected to autonomously navigate to and destroy a ground-based target.

The requirement was unrealistic for one reason: time. A satellite-guided bomb needs time to stabilize after release, power up, connect to satellites, determine a location, calculate directions, and begin navigating. Time was not available to an object traveling near the speed of sound straight toward the ground. The math was undisputable and the requirement unachievable. The 10,000-foot release requirement from a nose-diving fighter jet needed to change.

The debate between the product design team and the customer lasted months. It was finally resolved by the Chief of Staff of the United States Air Force, the military general officer at the top of the pyramid. He said:

> *"Anyone who uses GPS (global positioning system) guided munitions knows it takes... a good thirty- or forty-seconds time of flight to get the full accuracy out of GPS guided weapons. So you're not going to be down below 15,000 feet with properly released GPS weapons."*[9]

The warfighting customer heard their boss loud and clear. The unachievable 10,000-foot nose-diving requirement was changed. The new requirement was to release the bomb at or above 15,000-feet above-ground-level from an F-15 Eagle flying straight and level. The 15,000-foot requirement was achievable, and it opened the door for the team to leverage existing technology which sped up the design and development process significantly. No new inventions were required. By demanding an achievable requirement, the Small Diameter Bomb program was able to deliver the needed capability on time and well under budget.

The ten-second camera rotation and 10,000-foot nosediving bomb release requirements were both stable yet unattainable. The traditional program manager accepted the unachievable requirement and failed to deliver. The maverick demanded an achievable requirement, leveraged existing technology, and successfully delivered the product on time and under budget. The bottom line is project management professionals must demand achievable requirements to deliver fast, affordable, and effective capability. Achievable requirements allow the project team to leverage existing technology and keep inventions where they belong: in a laboratory setting.

Another key step highly effective program managers take to capture requirements discipline is avoiding scope creep. Scope creep is a causal factor in almost every product development failure. The Air Force relearned this lesson on a cancelled one-billion-dollar software system. That's right, one billion dollars was spent but nothing was delivered. Why? A new requirement was added to the software systems capabilities document daily. Yes daily. Simply stated, scope creep like this destroys project plans and kills programs. Whenever requirements control and discipline are lacking, this budget-draining and schedule-elongating disease will set in.

Although scope creep can rear its ugly head anywhere in the development process, two areas, in particular, must be properly controlled: requirements flow down and test. The requirements flow-down process takes the customer's "what we want" and breaks it into numerous "how to build it" product specifications. If the customer wants a hammer to pound nails into wood, the flow-down process would specify the hammer's size, weight, length, diameter, materials, and so forth.

The requirements flow-down process is ripe for scope creep. The F-22 Raptor training system upgrade team experienced this firsthand. The contractor hired to work the training system upgrade unexpectedly declared a three-million-dollar budget overrun. Instead of opening the checkbook, the program leader challenged his very capable Air Force project manager. "Find a way to deliver the upgrade on time and on budget," he said.

A few weeks later the talented training system project manager walked into the F-22 program leader's office smiling from ear to ear. "We solved it," he said. He went on to explain the government's requirements were flowed down by the contractor into many nice-to-have but not needed lower-level specifications. By diving into the details, the F-22 training system team identified and removed three million dollars' worth of scope creeping "desirements" being masked as specifications. The training system upgrade was delivered on time and on cost as a result.

Scope creep can infect a program during project testing too. In the final throws of testing an Air Force enterprise financial software system, the user continually identified test deficiencies requiring correction; or so it seemed. The program team eventually found the case-in-point. A system software function was tested and passed, but the customer's test agent "didn't like it" and wrote up a deficiency report. In the description, the test agent identified what they wanted the functionality to be, which was different than the original requirement. The software coding team reworked the functionality to clear the discrepancy, but the test agent failed it again. The test agent documented the second test "failure" with a note: "Return to the original configuration." That's right, a function passed the test but was written up as a failure, twice. The point is, the customer's test agents were failing test points because they did not like the requirement. This is known as gold-plating requirements, a sinister form of scope creep. The test agents were scope creeping requirements and it was happening at the end of product development, not the beginning.

Scope or requirements creep can sneak into any program, from any direction, at any time. This is why effective requirements control and discipline are required from the start of a program to the end. Effective program teams implement requirements discipline by demanding achievable requirements from the start and controlling scope creep until the end of a project. If they do not implement requirements discipline, their project slides down the slippery slopes of inventing or scope creep and predictably suffer cost increases and schedule delays.

An effective technique for implementing requirements discipline is to establish a Requirements Control Board. The control board is comprised of relevant stakeholders (e.g., customer, financing, testing, engineering) and chartered to vet all requirements and requirements changes. This ensures changes are vetted across the product's entire life-cycle and keeps a single individual from creeping the requirements into unobtanium. Establishing value-added project governance puts up the barriers needed to keep the creeps out.

Project management professionals must establish and enforce requirements discipline from the beginning of a project to the end. At the onset, they demand achievable requirements to leverage existing technology and move inventions to laboratories. Then they take the extra step to establish requirements discipline across the products development life cycle to keep scope-creep on the sidelines, not the field. This key point is worth repeating. High-performing project managers demand achievable requirements to leverage existing technology, dodge inventing, and avoid scope creep.

Chapter Endnotes

[9] Audio visual recorded speech by General John Jumper at Air Command and Staff College. Circa 2001. Specific date Unknown.

Fourth: Capture Trade Space

Design stability is more important than requirements stability

A foreign country is purchasing an American remotely piloted (unmanned) aircraft system. The product development plan is to use the American aircraft, sensors, and ground stations while integrating the country's own attack missile onto the plane. This is a risky strategy involving two countries, multiple companies, and a never-before-integrated weapon. Adding to the risk, communication, and coordination must be accomplished by teams literally an ocean apart.

Several executive-level meetings were conducted early on to enhance teamwork and communications. During one of these meetings, a foreign three-star military general asked his American counterpart, "If you had a single piece of advice for me, what would it be?" Without hesitation, the American General said, "Give your team some trade space."

To make the point, the American asked his foreign visitor a question. "If the aircraft and sensors were ready to field, but your country's missile was not integrated, would you still use the unarmed plane for

43

reconnaissance and surveillance missions?" The foreign visitor's answer would be telling. A "Yes" would mean schedule was his country's focus. A "No" would mean performance was the country's priority. Either way, the project team would capture trade space, or the ability to make trade-offs between cost, schedule, and/or performance requirements. The foreign dignitary responded, "Ahhh" holding his answer until he could discuss the question with his team. However, the answer to the question never came; the project suffered cost increases and schedule delays as a result. As the saying goes, when everything's a priority, nothing is.

Trade space is essential in the project management profession. Setting a clear priority between cost, schedule, and performance allows the project manager to make trade-offs between the other two variables. Once this trade space opens up, the project team increases the probability of success while also sharing risks with enterprise stakeholders. It is good business. The ability to make trade-offs while designing and developing products is an absolute prerequisite for delivering successfully. Period.

In the 1980s the United States decided to invest in a radar-evading stealth aircraft. Stealth performance was the clear national priority, budget and schedule were secondary. This is why the nation's taxpayers invested thirty billion dollars to design and develop the F-22 Raptor stealth fighter. The F-22 program office traded cost and schedule to achieve stealth performance.

However, F-22 trade-space prioritization was not publicly acknowledged and caused significant frustration in the nation's capital. Eventually, Congress articulated that budget was the primary priority, after which the Air Force began making the non-essential performance trade-offs and concessions necessary to expedite delivery. For example, some specialized support equipment requirements were eliminated.

The F-22 shifting priority from performance to cost highlights a central reality concerning large-scale product developments in any industry. That is, cost and/or schedule will always be prioritized ahead of performance. Senior executives or a board of directors will reach a point of frustration if they deem product development is taking too much time and/or money. At some point, even your customer is going to say, "Good enough."

Executive leaderships focus on cost and schedule ahead of performance is a reality, whether spoken or unspoken. This unconsciously pushes program managers to over-inflate cost and schedule estimates at program inception. This, traditionalists say, is the only way to manage expectations and create the reserves necessary to mitigate unexpected problems or realized risks. Overinflating cost and schedule estimates at the start of a program is another reason "zero-risk" project planning persists. Unfortunately, this type of project planning is the norm for program managers inside the Defense Department. Once approved, the project managers' overly conservative budget becomes the actual cost prophecy.

Here's an example of this phenomenon. The Predator remotely-piloted reconnaissance aircraft was introduced into combat during America's response to the attacks of September 11, 2001. The unmanned aircraft's real-time videos and loiter-time provided the United States with unprecedented situational awareness of enemy activity on the ground below. However, Predator was unarmed. Operators could see our enemies planting roadside mines, but nothing could be done in response.

The Air Force wanted Predator armed. The program team's initial plan called for three years of product development at a total cost of thirty million dollars. The Chief of Staff of the United States Air Force was not happy with what he dubbed the "zero-risk" planning. In frustration the nation's top-ranking Air Force General directed:

"You have three months and three million dollars. Get it done!"[10]

The Chief's statement did more than motivate the team. It also precisely articulated the program's tradespace. In a sentence, the Chief was telling the team the priority is on cost and schedule, so trade-off performance and "get it done."

The Predator program team immediately delayed or eliminated time-consuming, costly, and unrelated reliability and maintainability performance requirements. Requirements the users were unwilling to trade off previously. The result: Predator was armed in three months for three million dollars. The point is clear. Trade-space is the key to rapidly delivering capability, sharing institutional risk, and consuming less corporate treasure.

One method for capturing trade-space is to demand achievable requirements as discussed earlier. Another is to flex requirements to leverage existing technology and avoid the law of marginal returns. The last is to defer or even eliminate requirements altogether. The fact is, the ability to make requirements trade-offs is key to accelerating the product development process. In fact, the notion requirements stability enhances product development outcomes is a myth.

Many customers and stakeholders are led to believe requirements stability avoids the sin of scope creep. This is simply untrue. The C-130 stealth requirement from the previous chapter was "stable" yet unachievable. Stable requirements are also susceptible to scope creep during tests and requirements flow down as discussed earlier as well. The reality is the only place for stability in requirements is in operational suitability and safety. All other requirements should remain flexible and used as trade-space. Maintaining requirements flexibility until the design is stable and technology mature is paramount.

Design stability, not requirements stability, accelerates development and captures successful outcomes because it minimizes change. Change, no matter how small, can ruin a program. Here's a case in point. Jet engine fan blades are much smaller versions of the blade in a common household fan. The angled blades move outside air into the jet engine so thrust can be produced. The fan blades are the visible portion of the jet engine, and as such, are susceptible to environmental damage from pebbles, birds, and more. When a blade is damaged, it must be removed and replaced. The Air Force has a lot of jet engines and they buy a lot of fan blades annually.

Decades ago, the manufacturer sought approval to change the type of ink being used to print the model and serial number on each fan blade. The new ink, the manufacturer said, was less expensive and the savings would be passed along to the Air Force. The Air Force determined the change did not impact fan blade form, fit, or function and the "minor" change was approved. A few years later, jet engine fan blades began corroding at record rates. An investigation revealed the new ink and fan blade material were incompatible. The new ink, it turned out, introduced fan blade corrosion. To save a penny, the Air Force ended up spending millions on new blades.

There is no such thing as a "minor" design change. Change of any kind, at any point during product development or sustainment, is a costly bet. A gamble frankly not worth the risk. Design stability minimizes costly and time-consuming engineering changes altogether. This is why design stability, not requirements stability, is essential to improving acquisition outcomes.

The Air Force's Small Diameter Bomb program team implemented this principle skillfully. The program's competitive business strategy was simple. Prove, using data, that your design is more likely to deliver the needed capability on time and on cost, and your company will win the lifetime contract. This simple construct forced the two competing companies to adhere to the following logic pattern.

To increase the likelihood of delivering on time and on cost, reduce the program's cost, schedule, and performance risks.

To reduce risk, build and test components and/or the full product.

To build and test, complete and stabilize a design.

To complete and stabilize a design, leverage existing technology.

To leverage existing technology, make sure requirements are achievable.

The Small Diameter Bomb program office went one step further. They required competing companies to submit a bill of material in the winner-take-all final source selection proposals. This meant competing companies had to submit a list of the pieces, parts, and components used to construct the bomb they tested during the competitive design phase, and the list of materials they planned to use in the single-source product development and production phases. The bomb's source selection criteria stipulated deviations between the design and development material lists increased the vendors' risk and decreased the likelihood they would deliver the product on time and on cost.

The competitive strategy and risk-based selection process combined to make design stability the most important factor required to win a lifetime's worth of revenues and associated profits.

The simple business strategy worked. Both competitors completed critical design review, the bellwether of design stability, prior to the official program of record initiation. This is highly uncommon. Critical Design Review is normally completed much later in the process, during the subsequent product development phase. The design was so mature and stable, that the program went on to complete product development in under eighteen months at nearly four hundred million dollars below budget. The reality is, design stability is much more important than requirements stability.

There is an irrefutable truth in the product development business: late-breaking design or engineering changes break project budgets and schedules. For this reason, high-performing project managers place design stability ahead of requirements stability. By doing so they capture performance trade-space while keeping the focus on what matters most to their executive stakeholders: cost and schedule. The bottom line is capturing customer agreement to manipulate, defer, or eliminate requirements provides the trade-space required to deliver on time and on cost.

This key point is worth making again and again. Capturing trade-space accelerates product delivery and encourages all enterprise stakeholders to share the programmatic risk. *Design stability, not requirements stability, is a cornerstone of product development success.*

Chapter Endnotes

[10] Audio visual recorded speech by General John Jumper at Air Command and Staff College. Circa 2001. Date Unknown.

Fifth: Leverage Competition

Companies want to win a lifetime worth of profits; avoid paper promises, make them compete for it

The B-52 Stratofortress was initially delivered to the United States Air Force in 1962. The bomber aircraft is still flying today, sixty years later. The list of upgrades and modifications made on this reliable old jet could fill many pages of this book. The latest, a radar modification program, is bringing over two billion dollars of revenue to the original equipment manufacturer. The radar upgrade is the latest in a long line of billion-dollar upgrades the aircraft has received over its long and illustrious lifetime.

When a company wins a military product development contract, they win much more. They also secure future options for highly lucrative sole-source follow-on production, sustainment, and modernization contracts. Winning a competition to develop a military machine of war literally brings a lifetime's worth of profits to the winning company. This is one reason why every product development plan should include competition through at least the design phase. Make companies build and test hardware to prove-out their design. Avoid paper promises masking as competition; make companies compete.

People run faster when racing and so do companies. Especially when a company is chasing a product development contract worth millions or billions of dollars. As discussed in the previous chapter, the Small Diameter Bomb program understood and used this inherent motivation to record-breaking success. Their competitive design phase and "prove it via test" business plan resulted in unprecedented design maturity and risk reduction before the sole-source product development effort officially began. This is how the program was able to secure a fixed-price production contract and reduce the price per weapon by over fifty percent below even the most optimistic cost estimates. Government auditors estimated both competing companies spent two dollars of corporate treasure for every dollar of government money they received during the competition. With a lifetime worth of profits from production, upgrades, and modifications on the line, the C-suite executives were willing to invest.

On the other end of the source selection spectrum is the "paper promise" competition. Under the banner of this pseudo-competition, the winner is picked based on a written description of their proposed design. No evidence, no test data, and no proof. Just a paper promise. Contract awards under this mock competition provide no incentive for a company to invest its own money to prove their design is superior. There is no pressure to reduce costs or expedite schedules. An even worse outcome from this type of fake competition, is the incentives between the buyer and seller become polarized. A cost or budgetary overrun is viewed as a failure on the government's side of the relationship. On the sole source company's side, to the shareholder's delight, the overrun is considered "unexpected quarterly profits." Dueling incentives is no way to manage a marriage. But that is exactly what you get for marrying a vendor based on a written description. As the old saying goes, pick your partner carefully, because this single decision drives a high percentage of lifetime happiness or misery.

The deliberation between a true horse race and a paper-promise competition happens all too frequently. Even after the Small Diameter Bomb's highly successful dual-source competition, the paper-promise strategy option appeared again. This time on the follow-on variant of the weapon, called Storm Breaker. Like the original Small Diameter Bomb, Storm Breaker was required to glide to and attack stationary targets. Storm Breaker was also required to find, fix, track, and attack a moving target too. This meant the upgraded bomb required the eyes of an all-weather seeker.

Since Storm Breaker's primary requirement was to attack a moving target, the government's Small Diameter Bomb program team was replaced by more seeker-savvy personnel. Only a handful of Small Diameter Bomb program veterans remained on the Storm Breaker team for corporate knowledge purposes.

The newly assembled team's first task was to develop a product development strategy for Storm Breaker. The veterans favored a dual-source horse race competition. After all, the strategy delivered the original Small Diameter Bomb on time well below budget. The new seeker-savvy cadre favored a paper-promise competition. "Every dollar spent on a losing team is a dollar wasted," they said. The inference was the dollar spent on the losing vendor was a dollar the winning vendor could have used to expedite the product design and development process.

The debate did not last long. The new leadership team decided it was in the government's best interest to pick a single winner. A paper-promise product development plan emerged. The winner of a lifetime's worth of profits and revenue would boil down to who wrote a better proposal. No concrete evidence, just words on paper.

A few months later the Storm Breaker team had three single-source strategy options fully vetted and complete. It was time to head to the Pentagon for a decision to proceed. The night before the Pentagon meeting, however, the phone rang. On the line was the Air Force's Program Executive Officer for Weapons. The Program Executive Officer is the person the Pentagon holds accountable for delivering on time and on budget. Since the Program Executive Officer had significant skin in the game, she had a few remaining questions about the Strom Breaker strategy.

The program manager described the three single-source strategy options in great detail. Afterward, the Program Executive Officer asked the program manager a single question.

"If you are asked which of the three options you recommend tomorrow, what would you answer?"

The Storm Breaker Program Manager hesitated but eventually identified his preference between the three paper-promise strategy options. The weapons executive's response was unexpected at such a late date.

"Well, that would be embarrassing. I plan to recommend the dual-source competitive design phase strategy the Small Diameter Bomb program used so successfully. Add that in as an additional strategy option for tomorrow's decision meeting."

The next day the Air Force's Service Acquisition Executive selected a dual-source competitive design phase strategy for Storm Breaker.

Seven months later two companies were selected to compete during the Storm Breaker design phase. At stake, were a lifetime worth of Storm Breaker revenue and profits. The first vendor was the incumbent, the Small Diameter Bomb original manufacturer. The second company was new and late to this particular game.

The odds highly favored the original Small Diameter Bomb manufacturer. After all, they had nearly a ten-year head start, a mature design, proven technology, and a fully fielded small-diameter weapon. Their weapon was already gliding to and attacking stationary targets in a combat zone. Their contender was starting from scratch with one advantage: they had a proven air-to-air moving target seeker.

A few short years later the competition was over and the winner was announced. The hungry upstart outpaced the incumbent, proved their design was superior with test data, and won a lifetime's worth of work. Competition matters.

Critics of this principle will attempt to take an easy way out. They will argue they inherited a vendor-lock situation and are unable to compete. If any company has this type of sole-source lock, it is the F-22 Raptor manufacturer. This company truly possesses unique intellectual and proprietary radar-evading technology and know-how. The government's F-22 program team decided to compete anyway. They looked for the tasks being performed on the sustainment contract where proprietary knowledge was not required. One area they found ripe for competition was the contractor maintenance teams, known as Field Service Teams. These teams provide additional support to the Air Force organic maintainers at each operational F-22 base. The government team found the prime vendor was hiring former Air Force maintainers to do the work. Other companies could do the same, that is hire former Air Force maintainers to do the Field Service Team taskings. The Field Service Team tasks were broken out from the prime vendor, competed via small businesses, and costs dramatically decreased as a result.

Critics will also say a reason they can't break vendor lock and compete is because the prime contractor has unbreakable proprietary data rights. The F-22 program team challenged this myth too. They asked for a price for the data needed to conduct in-house or organic repair work on ten peculiar support equipment items. The cost to exercise the government's general-purpose data rights was extremely inexpensive. Today the government organically supports numerous items at a cost sixty-percent lower than the original equipment manufacturer charged.

Vendor-lock can be broken even for programs years into production. This was the case for the jet engines powering the Air Force's F-15 Eagle and F-16 Fighting Falcon jet fighters. The sole-source jet engine manufacturer developed a monopolistic mindset and engine costs ballooned and schedules delayed as a result. The Air Force's request to lower costs and maintain schedules went unheeded for years. In frustration and at great expense, the Air Force opened an alternate jet engine manufacturing capability. The dual-sources for F-15 Eagle and F-16 Fighting Falcon jet engines returned cost and schedule discipline to manufacturing.[11] The point is, regardless of the phase of the product development process, competition will speed timelines and lower costs.

The B-52 Stratofortress is sixty-years old. The F-16 Fighting Falcon is forty-five. The F-15 Eagle is forty-seven. Each aircraft has required decades worth of supply parts, upgrades, and modifications to remain safe, capable, and lethal. The original manufacturers of these aircraft won billions of dollars in revenue and associated profits well after each plane was produced. This lifetime incentive is why every project manager must leverage competition to speed up timelines and lower costs. At a minimum, find a way to compete through the products critical design review and watch how efficient and effective companies become.

Highly experienced project management professionals recognize companies are highly motivated to win a lifetime's worth of post-production support and modification revenue. They use this knowledge to leverage true competition, speed up project timelines, and lower costs. The bottom line is, successful project managers reject paper-promises altogether. Instead, they leverage competition forcing vendors to "prove it" using data before deciding who will win a lifetime worth of revenue and associated profits.

Chapter Endnotes

[11] See *The Air Force and the Great Engine War* by Robert Drewes. The full citation is located in the bibliography.

Sixth: Keep Stakeholders Informed

Turn critics into advocates

A foreign partner agreed to spend nearly thirty-billion-dollars to purchase American F-15 Eagle fighter jets and support. Deliveries were set to begin three years later. A month before the initial F-15 was set to be delivered to the foreign partner the Americans announced a two-year delivery delay. Surprise! The reactive declaration and numerous "how did this happen" questions kept the F-15 foreign sales project managers busy for months. Even worse, the Air Force's program management community lost credibility in the eyes of executive stakeholders from two countries.

The program team and the executive stakeholders from both countries had a communication problem for several reasons. First, program information sharing was reactive only being disseminated when risks became reality. When executive stakeholders receive no heads-up, and are the last to know, you have a problem. Second, the information being shared was interesting and not relevant. The number of foreign pilots and maintainers trained in the English language is interesting. The status of airplanes in production is relevant. You simply cannot pilot without a plane. Third, the program lacked a process to routinely disseminate relevant information, they lacked good governance. Irrespective of the background situation, the answer to the "how did this happen" question always boils down to one phrase: poor communication. Ineffective communication with stakeholders turns them into project critics. Effectively communicating with stakeholders turns then into advocates.

Properly communicating with stakeholders can be challenging, especially in large matrixed or geographically dispersed organizations. One key stakeholder may control funding from Washington D.C., another may be validating testing in Florida, while a different person is managing human resources from Texas. If this is not enough, add in the fact executive stakeholders have both formal and informal advisors whose tenure and history are persuasive. Nonetheless, there are several things effective program managers do to keep stakeholders informed.

The first and most important rule for keeping stakeholders informed is "no surprises." Nothing will diminish trust and confidence or disrupt a process more than a surprised stakeholder; nothing. Surprises make executive stakeholders waver, and wavering executives send in investigators and auditors to help them decide when to hit the program pause, rewind, or delete button. Surprises are an immensely disruptive force in product development. Focusing the program team on a "no surprises" theme opens the lines of communication from the bottom to the top of an organization.

Recall earlier in this book the program team told the F-22 customer "no" to a late-breaking Automatic

Backup Oxygen System requirements change. Unpopular decisions like this move up the chain of command in the blink of an eye. Knowing this, the program leader immediately notified his superiors of the controversial decision. The F-22 executive leaders agreed with the reasoning and provided both the authority and top cover needed to see it through. The bottom line is if you want your boss to have your back, you need to have theirs too. A highly effective program leader employs a "no surprises" rule for stakeholder engagement.

The second rule for effective stakeholder communication is to proactively manage the schedule. Stakeholders recognize there is no way to predict, plan, or schedule a product development effort with near certainty. They are forgiving as long as the program team is proactively managing schedule and keeping them informed. Team credibility will suffer, however, if you fail to proactively manage the program's schedule.

The delayed F-15 jet foreign sale mentioned earlier became an international issue because of reactive schedule management. This is what happens when program teams either abdicate their important oversight responsibilities or worse, develop apathy toward schedule management. Effective program managers work to hold everyone in the enterprise positively accountable for delivering on schedule commitments. They follow up to make sure tasks are starting and ending on time. This is the only way to proactively identify and mitigate schedule risks before they manifest into problems. Routinely communicating problems up the chain of supervision eliminates surprises, even when risks materialize. Reactive schedule management is not a personnel or process issue, it is a leadership issue. Leaders proactively manage schedules. This key principle is covered in more detail in the chapter titled "meet commitments."

The third way to ensure stakeholders are kept informed is to communicate relevant, but not interesting information. When a program leader shares too little relevant information, or too much interesting information, credibility suffers. Your key stakeholders need to know about the program's cost, schedule, and performance concerns. They do not need to know about the program's teambuilding picnic. As a Pentagon mentor told me, "Passion is interesting but only data is relevant."

Differentiating between interesting and relevant is a skill all leaders must possess in the complex product development profession. An often-told snowball story works to make this key point. After a fresh snow four brothers are tasked to shovel the driveway of their family home. After the chore is complete, it's time for some fun. The brothers pair up two-to-a-side of the driveway and tuck behind the large mounds of snow on either side nearest the road. After a few minutes making snowballs, it is 3-2-1 game on.

When the snowball fight begins one team executes a brilliant two-part plan. First, they throw a few snowballs high in the air. Second, while their blood-line opponents look skyward, the snipers sneak to the sides of their mound and throw line-drive snow balls at their distracted brothers for the win. The high arching aerial bomb is interesting, the incoming line-drive snow bombs are relevant.[12]

The program reviews at the Air Force's Foreign Military Sales office were interesting. The team covered the size of the foreign partner's military, number of planes, political climate, and last visits the American's had in-country. The team was silent on sales cycle-time, delivery schedules, project costs, risks, or issues. In fact, the reviews did not cover foreign sales in any fashion. The relevant data was missing. No surprise then, the foreign partners began asking, "Why is it taking so long and costing so much?"

The fourth way to keep your stakeholders informed is to establish good governance. These include recurring communication forums like working groups, oversight councils, and/or executive steering groups. This is especially true for programs with decision making conglomerates (e.g., joint military or multi-company partnerships), numerous customers (e.g., information technology programs), or involving multiple bureaucratic departments or agencies (e.g., international sales). In these types of demanding stakeholder environments, it is virtually impossible to effectively communicate individually with stakeholders. Complex stakeholder environments like these require the establishment of good governance structures to enhance communication. Good governance is the best way to keep a wide-spread group of stakeholders strategically aligned, plans integrated, and actions synchronized. Said another way, getting a diverse group of stakeholders to row the boat together requires them to be at the same lake and in the same boat.

The Air Force's Foreign Military Sales enterprise is a complex environment including the Departments of State and Defense as well as Congress. Making a tough environment tougher, an airplane sale involves nine of the ten Air Force Program Executive Officers (e.g., one executive's responsible for the engine, another the bombs).

During a F-15 fighter jet program management review, a foreign military general made a point masked as a question.

> *"Thank you for delivering the F-15 avionics test stand on-time. What should we do with it? The building it is going to be in, hasn't been built yet."*

The foreign sales enterprise was not rowing together: they were not aligned, integrated, or synchronized. To correct the situation two executive-level forums were created. The first brought the ten Air Force program executive officers together quarterly. The second brought the Air Force's numerous Pentagon foreign sales stakeholders together semi-annually. Communication improved dramatically and so did performance: requirements to sale cycle time reduced forty percent in three years.

The Air Force's Small Diameter Bomb and Storm Breaker programs also implemented effective stakeholder governance. They established a Joint Requirements Steering Group to vet numerous demands from their United States Navy, Marine Corps, and Air Force customers. This was the only way to control scope-creep and to ensure requirements were mutually beneficial and supported by all parties.

Effective governance is not "as needed." It is recurring with a well-documented purpose, a list of participating members, and the decision maker(s) clearly identified. The governance charter should also define the "we are done meeting when" criteria.

Despite your best efforts to limit stakeholder surprises, they will occur from time to time. The fifth principle to remember when communicating with stakeholders is to control your emotions, and not let your emotions control you. Remember, the goal of communication is to keep stakeholders informed. They are allowed, even encouraged, to disagree with and debate the program team. Let the process play out like a game of ping-pong where the stakeholders are the paddles, and the program the ball. Remember, questions are not a personal affront to the program manager's integrity or competence; they are a healthy part of the decision-making process. Healthy, honest, and unemotional debate eliminates groupthink and leads to better decision making.

The program leader for an enterprise software development program was glad to have this particular tool in the toolbox during the run-up to an important executive level decision. Months before the decision to field the software system was set to be made in the Pentagon, a major roadblock appeared. Two powerful advisors to the Pentagon decision maker, the Chief Information Officer equivalent in the Pentagon, were trapped in a policy dispute. One Pentagon advisor felt the program team needed to treat the pending decision like an upgrade, and leverage existing statutorily required documentation. By doing so, the Pentagon's review, coordination, and approval process would be expedited. The other Pentagon advisor disagreed and felt the program team needed to create a brand-new set of statutorily required documentation. Creating a brand-new set of program documents would take months if not a year. The software deployment would be delayed as a result.

The debate between the two powerful Pentagon C-suite advisors created a barrier the program was stuck behind. The project manager attempted to break the logjam by requesting, and was granted, a lunchtime one-on-one meeting with the Chief Information Officer equivalent in the Pentagon.

On the scheduled date and time, the project manager travelled five-hundred miles to the decision authority's office only to be told the meeting purpose had changed. Instead of being a thirty-minute office call to discuss program documentation, a full C-suite level executive board meeting was being convened three-hours later. One of the two Pentagon advisors convinced the C-suite executive the Air Force's program manager was being quarrelsome and not following direction from above. The project manager felt ambushed for sure, but had the emotional intelligence needed to keep the upcoming meeting impersonal and fact-based.

Shortly after the lunchtime hour the program manager was in front of a room full of executive stakeholders. Rather than call out the Pentagon advisors, the roadblock, or their ongoing debate, the project manager simply asked a question.

"Should the program leverage existing documentation and stay on schedule, or author and staff entirely new documentation and delay fielding?"

On queue the two Pentagon C-suite advisors began making their case disguised as questions. Rather than answer, the program leader said nothing. Soon the two Pentagon advisors began openly debating their own questions. The stakeholders in the room, including the C-suite decision maker, moved their gazes onto the two powerful executive debaters. With everyone's attention focused on the executive debate, the project manager slowly left the podium and took a seat along a side wall in the large conference room. The project manager's point was clear: this debate has nothing to do with me or the program, it is a policy dispute that requires resolution.

The Chief Information Officer equivalent listened to the debate for a minute, then looked to the vacated spot at the front of the conference room where the project manager once stood. Scanning the room, she finally found the gaze of the project manager. "I get your point," the decision authority said, "I have a policy issue to resolve for you." A week later the decision was made and the software program deployed on-time as planned. An effective project manager captures the personal maturity required to control their emotions instead of letting their emotions control them. A famous movie quote summarizes this point best: it's not personal, it's business.

Project management professionals receive their direction and resources from executive leaders. Placing a foot on the program accelerator requires effective and efficient communication. Keep stakeholders informed by limiting surprises, communicating relevant information, proactively managing schedules, establishing good governance, and remaining professional in the heat of bureaucratic battle. This is how you turn even your staunchest critic into an advocate.

Chapter Endnotes

[12] This story is found in my book, *Hustle and Have Fun! A Coach's Guide to Winning Over Players and Parents.* The full citation is located in the bibliography.

Seventh: Articulate Your Intent

Empower the team at levels

The United States Army's Command and General Staff College is located at Fort Leavenworth, Kansas. The school's military students refer to it affectionately as the short course; the military prisoners take the long one. The school's Commanding General, the Commandant, took the stage in the year 2002 to welcome a new class of over one-thousand students to their year-long military education.

> *"Welcome to the Command and General Staff College. You are here to learn from our incredible faculty and each other. I know you will. But if you do not also reconnect with your families while you are here, you are doing something wrong."*

The general's message proved prophetic. Many of the students in the auditorium previously spent time in Afghanistan responding to the attacks of September 11, 2001. These same men and women would later learn they were headed back to the Middle East after graduation, this time for Operation Iraqi Freedom. The year of no-deployments family time was precious indeed.

The commandant's succinct, unambiguous, and purposeful message is known as the Commander's Intent in the United States Army. This end-state-focused statement motivates initiative-taking and maintains unity of effort and purpose during the confusion and chaos of battle.

The first known product development effort to use this technique to great success was the Small Diameter Bomb program. The Program Intent truly came from the top of the pyramid, from the Chief of Staff of the United States Air Force himself. Getting the intent statement from the Chief wasn't intentional. It was out of frustration.

As seems to happen too often in the product development business, the Small Diameter Bomb customers prioritized cost, schedule, and performance equally. The iron triangle, as it is sometimes called, denied the program team the necessary trade-space required to share intuitional risk and make logical trade-offs. The trade-offs necessary to deliver the needed capability on time and on cost.

Despite the program director and team's best efforts, the customer would not budge. Cost, schedule, and performance remained equally important; no trade-space. In frustration, the program director phoned a friend at the Pentagon to seek some guidance and mentoring. Unbeknownst to the program director, his friend had moved to a new job; he was now working directly for the Air Force Chief of Staff. The discussion soon turned from catching up to business when the director's Pentagon friend asked, "How can

I help?" The Small Diameter Bomb Program Director replied, "I'll send you something to consider."

A month later the "something" became a short paragraph signed by the Chief of Staff of the United States Air Force, the Air Force's Chief Executive Officer.

> *"You will deliver Small Diameter Bomb in Fiscal Year 2006. Schedule is the priority. Initial delivery is for the F-15, but make no mistake, it must also be ready for the F/A-22."*[13]

Everyone associated with the program, from design to test, now knew the program's schedule-driven end-state. The program intent statement also captured the trade-space needed to get there. But there was one more equally important and intended consequence of the program intent statement: empowerment.

Program personnel were empowered at all levels to make the decisions required to deliver the capability in 2006. The program director reiterated this point to the team often: "If it helps us meet the Chief's intent, don't wait on me, do it!" No mother-may-I, no excuses, just a 'get the job done' focus; all from a simple intent statement. A common program intent empowers and unifies a team. It also enables them to avoid costly and time-consuming indecision, or worse, analysis paralysis.

Empowerment is a powerful force multiplier, but it does come with risk. There is an inherent assumption all teammates are ready, willing, and able to be empowered. Sometimes they are not. It is a leadership responsibility to coach your teammates so they can develop, learn, and grow. And once they have, turn them loose. If you have coached them well, they will help the team succeed.

If you feel you do not have the organizational authority to create a program intent, identify who does. Then ask them to define the program's end-state and priorities. This will get the conversation started and begin focusing, integrating, and synchronizing the efforts of the entire team.

Storm Breaker, the follow-on to Small Diameter Bomb, did not have the same top-level support as its parent at program initiation. Instead, the program manager created a "Program Manager's Intent" statement. During a briefing to his supervisor, the manager asked, "do you agree with this intent?" The supervisor did, so the intent slide was renamed, "Program Director's Intent." The same process followed two-additional times. Soon the intent statement was powerfully titled "Assistant Secretary of the Air Force for Acquisition's Intent."

Portfolio managers, especially those with geographically separated teams, will capture similar benefits by using a program intent. An Air Force information technology portfolio manager, for example, had five simple themes for the team's seven hundred personnel located in five states.

- *Take care of people*
- *Implement requirements discipline*
- *Keep stakeholders informed*
- *Deliver capability on time / on cost*
- *Continuously improve*

The portfolio manager then overtly communicated a simple strategic theme. "If it helps us take care of people, implement requirements discipline, deliver capability on-time and on-cost, unify stakeholders, or improve; do it." The team was empowered to do just that and ended up with a record-setting ninety-six-percent on-time delivery rate for the hundreds of software projects and upgrades in the portfolio.

Program intent statements are powerful. They provide the entire enterprise team, from project members to stakeholders, with a common focus, end-state, and priorities. Armed with guidance and support from the top, the team is empowered to make the micro-level decisions required to satisfy the intent. This may appear small, but a significant benefit is realized. Decision-making at all levels is sped up, and by doing so, product design and development is expedited too. Rapidly delivering products requires an empowered team at all levels. A project or program intent statement is a powerful force multiplier.

Indecision can create analysis paralysis just as fast as single-entity centralized decision-making will. Expedited decision-making, however, has the opposite effect. Effective project managers who earn a reputation for filling customers' needs rapidly use empowerment as a catalyst. The fastest way to get the entire team on the same page is by articulating an unambiguous program (or portfolio) intent statement. This will define the team's common purpose, end-state, and provide the guardrails for empowering the team at all levels. A succinct program intent is how high-performing project managers create unity of purpose, articulate the trade-space, and empower their team-mates so everyone in the boat rows together.

Chapter Endnotes

---

[13] In an attempt to show the versatility of the F-22, the Department of the Air Force briefly designated the jet a Fighter-Attack (F/A) aircraft.

Eighth: Be Decisive

In the absence of leadership, assume it

The Mine Resistant Ambush Protected Vehicle (MRAP) is designed to protect our soldiers from hidden roadside explosive devices. Delivering MRAPs to the field was a national priority in the mid-2000s. The need was so urgent the Secretary of Defense received regular Saturday morning program updates.

One Saturday morning the Secretary of Defense expressed displeasure with the MRAP delivery schedule. "I want more MRAPs delivered to Iraq and Afghanistan this year," the Secretary said.

The Marine program director reminded the Secretary of Defense MRAP's budget was locked; Congress had already authorized and appropriated the yearly funds. "What about this year don't you understand, Marine!?" the Secretary shot back. The Marine program manager smartly took the direction and requested a week to work out the details.

The next weekend the program director laid out the plan.

"Sir, you challenged us to deliver more MRAPs to Iraq and Afghanistan this fiscal year. We can reprogram funding from the spare parts account and use it to deliver more MRAPs this year instead. We will need your help to convince Congress to approve this funding reprogramming action though."

The Secretary of Defense's "do it" was met with a Marine program director warning.

"Sir, there is a risk you need to know about. We are going to take all the money from the spare parts account to do this. We could reach a point where there are not enough spare parts to keep the MRAPs already in the field up and running."

The Secretary of Defense responded, "I understand Marine, I accept the risk. Now get it done!" With that the meeting was over.

About ten months later the national media was reporting the fact MRAPs were sitting idle in the Middle East. The implication was the dirty rotten Marines inadequately funded the MRAP spare parts pipeline. There was no mention of the Secretary of Defense, the Saturday morning meetings, or who accepted the supply chain risk. An effective program manager recognizes whether they want it or not, whether its perceived or real, they own all the risk for their program.

The Expeditionary Combat Support System was intended to replace two-hundred-plus Air Force logistical information technology systems. The enterprise-wide system experienced severe scope creep and the associated cost and schedule problems were unyielding. The well-advertised program problems landed the system on a list of the twenty-six worst-run information technology programs in the Federal Government.

The honor cost the Expeditionary Combat Support System Program Director his job. The same fate befell the second program director. Customer-driven requirements creep sunk both program directors' careers, but not the scope-creeping requirement's officers. The third program director kept his job because the Air Force needed someone to kill the program, terminate the contracts, and pay outstanding bills.

Fair or not, when it is time to hold someone accountable for product development failure, it will be the program manager. C-suite leaders, political appointees, the Pentagon, and even the customers are never implicated, ever. An effective project manager recognizes this fact and makes the decisions required to achieve success. They lead.

The Air Force's Foreign Military Sales portfolio had a four-million-dollar problem. The enterprise's undisciplined hiring process brought too many people onto the payroll. The extra personnel broke the budget and someone needed to go to the Pentagon and convince them to pay the four-million-dollar personnel bill. A "what are we going to do about this" meeting was convened by the headquarters personnel and finance leads. The foreign sales portfolio leader took a seat and as soon as the meeting commenced, asked, "Who's in charge?" The question was intended to determine who was going to take the responsibility to fix the four-million-dollar problem. The silence was deafening.

Although the foreign sales portfolio lead had no control over hiring or payroll, he knew if the system was going to hold someone accountable for the four-million-dollar mess it would be him. "Okay, I got it," the portfolio lead said breaking the uncomfortable silence. After all, if you are going to take the blame for a predicament, assume leadership and make the decisions needed to fix it!

The foreign sales team identified the hiring issue's root causes, made changes, and fixed the problem. This satisfied the Pentagon and they released the extra personnel funding. A "one-time good deal," the Pentagon said. That's all the foreign sales team needed; the resourcing problems were resolved.

The "who's in charge" question should be asked much more frequently in the project management profession. Organizational responsibilities are oftentimes so widespread it is almost impossible to hold anyone even remotely accountable. But as the MRAP and Expeditionary Combat Support System examples highlight, the bureaucracy will hold someone accountable. Nine out of ten times, that will be the project manager.

High-performing project managers don't put their programs or careers in the hands of others. Instead, they follow a simple rule: in the absence of leadership, assume it. Afterall, if you are going to be held accountable for process dysfunctions, then make the bold decisions required to avoid the quicksand and succeed despite the bureaucracy. As the saying goes, it is better to make a decision and beg for forgiveness, than suffer cost, schedule, or performance issues while awaiting permissions. Effective project managers assume leadership from the start and take control of their program's decision space. They are decisive.

The "In the absence of leadership, assume it," principle is highly effective in the product development profession. Here's another example. A foreign military sales program manager announced a five-million-dollar increase in the price a foreign partner was going to pay for aircraft technical documentation support. The additional money was needed, a foreign dignitary was informed, to double the size of the American team responsible for the upkeep of their aircraft's support and maintenance documentation. When asked "Who made the decision?" the room fell silent. The foreign dignitary was quietly outraged too.

The American foreign sales portfolio manager put the "in the absence of leadership, assume it" rule to use after a few clarifying questions also went unanswered.

> *"I am going to make a decision which may not be mine to make. The United States is not going to raise the price for technical documentation management by five million dollars. We will continue to support our foreign partner the way we do it today and charge them the same prices."*

> *"Now I assume I am the wrong person to make this decision. If so, just ask the appropriate person to contact me. If the decision changes at that time, I'll notify the foreign partner. But for now, lacking any justification or knowledge on who made the decision in the first place, we will continue to do the work as is and not raise the price."*

Months later "who" made the original decision to increase the price was still unknown. To the great satisfaction of the foreign partners, the foreign sales portfolio leader's decision stood and charges remained the same.

In a large bureaucracy openly assuming leadership and making decisions may appear difficult. It is not. Well-intentioned people inside the bureaucracy appreciate decisiveness. They share the project managers' frustration with hand-wringing overly conservative bureaucrats. The bureaucrats themselves appreciate someone else jumping in and moving out. The point is, project managers should take the job very personal, because if the project goes south for any reason, the blame will land squarely on their shoulders. Take it personal and make the decisions needed to enable project success.

There are several major cautions on this particular principle. First, as you take the initiative to help your team succeed and satisfy your customers, keep all decisions in alignment with the project's approved goals. Second, make sure decisions are aligned with the organization's (and your own) values. Third, keep integrity as your moral compass. Finally, keep stakeholders informed so they are not surprised.

"Lead, follow, or get out of the way" is a commonly used phrase. When it comes to effective project managers, this saying is doctrine. This type of project professional does not "wait" for others to make decisions. They make the decisions required to enable project success and move out, especially when the "Who's in charge" question remains unclear. High-performers recognize extreme action is necessary at times to deliver rapidly for their customers. Be decisive; in the absence of leadership, assume it.

Ninth: Meet Commitments

Proactively manage schedules, build trust and confidence, remain credible

The Air Force Life Cycle Management Center Commander is responsible for armament, electronics, aircraft, and associated product development efforts for our nation's aerospace warriors. The center has 26,000 employees, but only a few hundred get the honor of being called program directors.

The theme for a two-day meeting for all of the center's program directors was "Changing the Culture of Air Force Product Development." A resurgent strategic competition with Russia and China makes moving from deliberate to rapid product development an absolute necessity.

The final topic of the conference was a discussion of the culture change captured in the Air Force's Foreign Military Sales enterprise. The foreign sales portfolio director covered his team's three-year journey; from the team's early resistance, to cautious overwatch, and finally active engagement. The pitch included ten improvements the team implemented to increase both their span of control and influence in the foreign sales enterprise. These included good governance, data-driven decision making, project baselining, and proactive schedule management. The results of the culture change spoke for themselves: a forty-percent sales cycle time reduction with customer and employee satisfaction at all-time highs.

The foreign sales portfolio lead ended the one-hour session with a project management lesson thirty-years in the making.

> *"As many of you know I am retiring soon. And I wondered, if I was to leave you with one thought to help you rapidly deliver the world's most feared and respected Airpower, what would it be?"*

> *"I'd say, read a Fortune magazine article titled, "Why CEO's Fail." I read this when it was first published in 1999 and it truly shaped my leadership style."*

> *"The short four-page article compares two groups of Chief Executive Officers: a group that kept their jobs by increasing stockholder value, and group that was fired for not doing so."*

> *"Both groups were smart, experienced, and crafted plans focused on increasing their respective company's valuation. For all these similarities, they had one major difference."*

> *"One group followed up to make sure their subordinate leaders were meeting their commitments and implementing the plans needed to grow the company. The other group failed to follow up altogether, and their subordinate leaders were not meeting their implementation commitments. The company's valuation suffered as a result."*

"As a program leader, you must follow up to make sure your team is setting and meeting their commitments. This is how you build trust, confidence, and credibility. More importantly, this is how you deliver timely, relevant, and lethal war-winning capabilities to our warfighters. I know you will deliver on your commitments because if you do not, our nation will lose the next conflict."

The cornerstone of my career as a program and portfolio leader was setting and meeting commitments. Make a promise, keep a promise. This is the only way to earn the trust and confidence of others and capture an important professional asset: credibility. Effective program management is a marathon, not a sprint.

Throughout my career, the attitude that frustrated me the most was a laissez-faire approach to schedule management. Missing even seemingly minor commitments creates downstream ripples everyone else has to accommodate. The government's failure to send out a Request for Proposal on time impacts the entire food chain. Potential vendors experience increased overhead costs as the bid and proposal staff has to be paid to wait. The government customer suffers because the needed capability is delayed as a result too. These are the bookends of the problems associated with a minor slip to the government's request for a proposal release date. There are thousands of these so-called "minor" schedule slips creating downstream tidal waves of delays and issues inside the Department of Defense's product development establishment. The examples of schedule mismanagement could fill the pages of a long novel. But just one example will do.

A military veteran opened a company supporting the Department of Defense upon his retirement from active duty. It was a rough start. He literally did not take a penny of profits from the company for his first five years in business. But the fifth year in business his company won a big contract to support the government. The celebration with his teammates, however, was premature. The government contract award was delayed for an unknown reason.

In a silent protest, the company owner decided not to shave until the contract was awarded as promised by the government. All that was needed after his company's proposal was fairly selected was a signature by a government contracting officer. Needless to say, his beard was full, long, and gray when the contract was finally awarded a full year later.

The inability to maintain schedule discipline is not isolated to the Government. A large jet engine company forwent eighty million dollars of revenue two years in a row. They could not get their proposals submitted to the government on time, and the Pentagon redistributed the funding as a result. Yes, a for-profit company lost one hundred sixty million dollars in revenue because they could not properly manage proposal due dates.

People in the enterprise are not evil, lazy, or incompetent. Schedule delays persist in a culture lacking accountability. "I have done my job, I'm waiting on," or "It is out of my control," or "I'm too busy to get to it," are common refrains. These are excuses marginal leaders accept, and by doing so you allow apathy to win out over accountability.

Missing deadlines, incomplete tasks, or delaying deliveries in product development is not a resource issue; it is a leadership issue. When these conditions are present, adequate planning and accountability are absent. This problem is easily identifiable. Program reviews normally include some form of schedule summary. If the tasks listed in the schedule summary are due six, twelve, or twenty-four months in the future, you have a schedule discipline problem. Period. It is easy to mask schedule mismanagement when the only commitments made are half a year or longer into the future. By then, nobody will remember who committed to what or when.

Reactive schedule management has become a disease. The F-22 Stealth Fighter program suffered it too. The two-hundred programs, projects, activities, and efforts under the F-22 umbrella include supply chain, new weapons integration, software upgrades, and more. More often than not, the team was failing to meet their task completion deadlines. When this happened, last year's work had to be accomplished using the current year's money. Current year tasks then had to be delayed. One late deadline created a ripple in the pond and the other one-hundred ninety-nine projects suffered.

Why was this happening? Because the teams were only tracking completion dates nine to twelve months in the future. When the day finally arrived, everyone was surprised to find the project was incomplete. The brick-by-brick tasks required to incrementally build the wall were simply not being developed, assigned, or tracked. It was like setting the wedding date without getting engaged or sending out the invitations. When the day finally arrives, you are standing solo at the altar reactively thinking, "Oh yeah, I forgot to do a few things."

The F-22 program team needed to move from reactive to proactive schedule management. The first step was to baseline all two-hundred projects. To do so, the teams were asked to complete a simple statement.

We will deliver _what_ by _when_ for _how much money._

This statement became the specific project's management baseline. Next, project managers were required to document the subordinate tasks required to meet the baseline commitment. This outlined the step-by-step tasks or actions required to achieve the "what." If a couple wants to get married by the end of next year, they need to select a venue, hire a caterer, send out invitations, get a marriage license, and so on. This is inch-stone planning.

The teams then completed backward scheduling on the inch-stone tasks. Keeping with our wedding analogy, the invitations need to be mailed out to the guests eight weeks before the wedding date. Mailing out the invitations on time requires the invitations to be selected and ordered twelve weeks prior to the nuptials. And so on.

After the F-22 project management teams completed defining the subordinate tasks and back-ward scheduling, it was time for a review. Project tasks more than ninety days apart required more subordinate tasks to be added. This process continued until all two-hundred-plus projects consuming taxpayer treasure were baselined with detailed task-oriented schedules. The task-oriented schedules were then consolidated on a single-page document for ease of use and tracking. A generic example of this simple project baseline and commitments tracker is located in an appendix.

The final step to recapture schedule discipline was to integrate all two-hundred task-oriented schedules on a single spreadsheet. This task-oriented F-22 Program Integrated Master Schedule was sorted by date, and reviewed weekly by the leadership team.

During the first weekly review, most project tasks were color-coded red for "late." But within three months almost every project task was "green" indicating completed on-time. Schedule accountability and proactive schedule management returned to the F-22 program office. Stakeholders trust and confidence soon followed. But just as important, by consistently meeting commitments, the F-22 team consumed less corporate treasure too.

Instilling disciplined schedule management has this effect across portfolios and complex enterprises as well. The Air Force's Foreign Military Sales process is highly complex with touch points across the Departments of State and Defense, as well as Congress. The process is purposefully cumbersome. After all, the United States does not want to give military technology to a country that would use it against us. The foreign partners, however, had grown tired of the sales delays and expressed frustration to the Secretary of Defense. The resultant direction from the Pentagon to "speed up" the sales process was unambiguous.

The foreign sales team expressed frustration at the new direction. The common retort was, "We did our work in five days" but are waiting on "International Affairs," or waiting on the "Defense Security Cooperation Agency." The waiting on excuses kept coming. When your only focus is on what you control, you forget how much you can influence.

Eventually, the foreign sales team embraced change and implemented a new form of governance and schedule discipline. Since the process included multiple disparate organizations, the team worked with the internal stakeholders to capture agreement on who would do what, and by when. The enterprise agreement was documented on a product nicknamed a "Focus Chart," so named because it focused numerous stakeholders on a common plan and end-state. The team color-coded Focus Chart tasks to highlight the responsible offices. For example, red indicated International Affairs at the Pentagon was accountable for the task, and blue indicated the Air Force's Security Assistance project manager owned the effort.

Finally, a series of recurring meetings or governance processes were implemented to bring the stakeholders together to review status. The foreign sales teams reviewed tasking monthly, and the executive stakeholders did the same quarterly. The forums identified issues, removed roadblocks, and changed policy as needed to bring positive accountability to the enterprise. Good governance is a highly effective technique for increasing the span of influence.

Bringing proactive schedule management to the foreign sales process resulted in a forty percent reduction in requirements-to-sales cycle time in just three years. Wow. A single integrated schedule, color coded by office of responsibility, reviewed frequently, with follow ups to accountable offices resulted in positive and lasting change. Following-up works.

There is a misperception rapid delivery of products in a large bureaucracy requires special permissions or secret processes. The reality is organizations known for rapid delivery of products have earned the trust and confidence of the bureaucracy by setting and meeting commitments. This trust translates into program advocacy where bureaucrats are willing to remove obstacles instead of placing them in the way. Highly effective project managers recognize this central truth, and work to proactively manage schedules, build system-level trust and confidence, and earn the thing that matters most: credibility. The foundation stone of effective project management and rapid product development is setting and meeting commitments.

Tenth: Find Your Why

Remain passionate about your higher-level purpose

The President of the United States travelled to Ohio to give a speech at the National Museum of the United States Air Force, attend a fundraising lunch, and visit with the region's Gold-Star families. Gold-Star is a name honoring the families of our nation's warriors who were killed in action. The families of our nation's selfless heroes; the ones who gave the ultimate sacrifice in defense of our way of life.

The President completed the first two items on his agenda and returned to Wright-Patterson Air Force Base for the most important appointments of his day. The agenda allocated one hour for the President to meet with twenty-six Gold Star families before he departed. But he paid little attention to the schedule. The visitations lasted over four hours. It was a humbling day for everyone involved.

The families began arriving an hour before the scheduled meeting times. As they did, an active-duty service member greeted and escorted them to a temporary Presidential lobby. "Hi, welcome to Headquarters Air Force Materiel Command," I said to receive another arriving family. "Is your whole party here?" The mother, standing with her four adult children, looked back toward the doors, and pointed.

"Oh no, he's back there," she said motioning to her husband whose tardiness was clearly frustrating her.

The Gold-Star father was lagging behind because he was determined to shake hands with every uniformed service member along his path. With each grasp, he passed a business card to the military member. True to form, when he approached me, he passed a business card with a handshake.

"God bless you, and thank you for your service," the father said. I thanked him, then placed the business card in my pants pocket without a glance, and led the family to the Presidential waiting room.

The President's staff secured four office suites in the military headquarters building to host the Gold-Star Families. The plan was for the President to spend fifteen-minutes with up to seven families at a time in each room. The one-hour timeline made it appear the President would enter a room, talk to the group as a whole for fifteen-minutes, then move on to the next room and do the same. But that's not what he did. Instead, the President visited and grieved with every family individually.

Once the last family was escorted out, the President's staff hustled the military escorts to a back stairwell. The President wanted to thank us; we were told.

"Wait here," the President's security detail said. A few minutes later the President appeared at the top of a twisted stairwell, and after seeing the uniformed military at the bottom, stopped and stepped backward up a few steps to disappear. After a brief pause to collect himself, he made his way down the flight of stairs and greeted the military escorts one by one. The weight of the world was visible. The President's eyes were reddened and his early-in-the-day television makeup tear smeared. His sincere empathy for our nation's lost heroes and their families was evident. But the sight of the uniformed military escorts seemed to give the president a momentary emotional boost.

"Colonel," the President said as he neared the end of the line of military escorts, "how's morale?" The answer satisfied the President, and he turned to greet me, the last escort in the line. As he turned and shook my hand, he seemed surprised. "Oh, another Colonel," he said and paused. I filled the awkward silence impulsively, "God Bless you Sir." The President paused again, then looked into my eyes and said, "Thank you, may God Bless us all." Then the President waved his photographer forward to take a group picture.

After the photo, the President said his rushed goodbyes as his security detail hurried him toward the exit door. But after thirty steps toward the exit, the President stopped and turned around.

"One last thing. Never forget how important you are to this nation or how much we appreciate the sacrifices you and your families make to defend our way of life. On behalf of a grateful nation, thank you."

The President turned back toward the door and departed.

I made my way back to my small cubicle desk in the headquarters' building. As I did, I felt a slight pinch in my pocket; it was the business card I received hours earlier from the father of five. The patriarch who stopped to briefly chat with everyone in uniform. I pulled out the card expecting it to be a business card. It wasn't.

May God Bless and Keep You

Proud Parents of Army Cpl Samuel Pearson Killed in Action, 10 Oct 2007, Baghdad Iraq

---

The emotions of the day caught up to me and tears began quietly streaming down my cheeks. Thoughts of children who will never hug their fathers, or mothers who will never hear "I love you" from their child again. Families torn apart to the point of having to be separated in two different rooms to meet the President. The Gold-Star Families exposed the true meanings of gratitude, compassion, and empathy. Of duty, selfless service, and love of country. Our nation's lost warriors are justifiably called heroes, and their Gold Star family members are too![14]

The Pearson's lost their son Samuel in combat. Despite their grief, they were thanking everyone in uniform for their service. They were thanking us for being Samuel's teammate, his brothers and sisters in arms, and his military family. They were grateful to those in uniform for honoring Samuel's sacrifice by tirelessly and continually defending our way of life.

Army Corporal Samuel Pearson and his family are one of the "why I served" our nation passionately for three-decades. So are A, J, L, and B. A and J died in aircraft accidents. L lost both legs during a road-side attack. And B suffers from post-traumatic-stress-disorder. I served to honor all of them, and to honor the nearly one million others who died to secure our freedoms. We owe it to all of them to bring a never-quit, no-excuses determination to our daily postings. We honor their sacrifice by doggedly executing the mission, and rapidly delivering the technological advantages our nations warriors need to fight, win, and return home safely.

A personal passion for your customer's needs brings out a sense of urgency and accountability unmatched by any other incentive. Find your personal why, the higher-level reason you do the job you do. Use it to fuel your daily fire to deliver for those you serve, and treat the job as if your own son or daughter was the customer. Army Corporal Samuel Pearson, his family, and the million who went before him deserve your absolute best effort every single day!

Chapter Endnotes

---

[14] A heartfelt thanks to the Pearson Family for their permission to share this very personal story.

Conclusion

The Mindset of High-Performing Project Managers

A new Deputy Director for the F-22 System Program Office was announced without fanfare. The inbound Deputy would be a Chief Operating Officer equivalent responsible for three-hundred personnel in five states. The deputy would help manage modernization and sustainment projects worth nearly two-billion-dollars annually. The F-22 Deputy Director position is a job every career civil servant would willingly embrace. Except for one.

I enthusiastically sent my newly announced deputy a congratulatory note, and received a single sentence reply: "I'm not coming!" We had met four-years earlier so I knew this was an in-character response. The previous meeting took place at the top-secret Big Safari Program Office in Dayton, Ohio. The "not coming" leader was the Big Safari Deputy Director at the time, and I was at the Pentagon justifying Big Safari's, and other intelligence programs, budgets to the Office of the Secretary of Defense and Congress.

The initial meeting was officially intended to discuss how the Pentagon could better support Big Safari. The Big Safari Deputy Director's focus, however, was to tell the budget puke from the Pentagon to "stay out of our way." To make his "your-not-one-of-us" point, he rolled out the "who's who" of Big Safari's impressive accomplishments. Slide after slide included media reports of rapidly fielded Big Safari systems; systems making significant contributions to our nation's warriors in Iraq and Afghanistan. I interrupted the Big Safari "stay out of our way" pitch about halfway through the scrapbook. "I get it," I told my host. He sniped back: "I doubt you do!" But he let me proceed.

I went on to say the Big Safari sales pitch was not about military systems, processes, or decision authorities. In effect, the deputy director wasn't talking about his impressive resume of Big Safari accomplishments at all. The story being told was about the culture of the Big Safari team and of the individual members lucky enough to be handpicked members of their exclusive club. The point the Big Safari Deputy was really making was, "You are not a member of the club, so leave us alone."

The Big Safari Deputy's poker face never wavered. "Go on," he sarcastically said. I proceeded to explain my perspective on The Big Safari Way. The "way" to rapidly deliver war-fighting capability on time and on budget. The mindset I had witnessed early in my career and used myself for decades as a program leader.

First, "go fast" is a deeply rooted tenet of The Big Safari Way. Anything that slows them down is frowned upon. They will not allow unachievable requirements or inventions to creep into their product development efforts. They strive to leverage existing technology and rapidly deliver products to the field. If they deem a requirement unobtainable, they change or scrap it altogether. Customers unwilling to maintain requirements flexibility are told to find someone else to do the work. Big Safari basically tells the entire establishment to get lost unless they agree with the product requirements.

Second, Big Safari never accepts a 'No' from someone in the bureaucracy who cannot tell them 'Yes.' Only the actual decision-maker's vote is relevant to the Big Safari crowd. Opinions from others are noted as interesting. This helps Big Safari avoid time-consuming and costly compromises while navigating or plowing down no-value-added bureaucratic roadblocks. In fact, if a staffer in the Pentagon tries to slow down their progress, Big Safari moves around the quicksand so fast the bureaucrat can't even react. Big Safari bypasses them and heads straight to the decision-maker's office. They are absolutely laser-focused on rapidly delivering capability to the warfighter and prioritize product ahead of (but not in lieu of) process every time.

Finally, Big Safari keeps their key stakeholders informed through a series of classified forums. The forums keep stakeholders informed by passing along relevant information to limit surprises and speed-up decision making. Their forums, in essence, turn stakeholders into program advocates who help Big Safari speed through or truncate the normal processes. They earned this advocacy due to a decades-long reputation of meeting programmatic commitments. They have earned the trust and confidence of the entire bureaucracy.

"That's it," I told my Big Safari host. "The Big Safari Way boils down to three key principles: deliver fast, don't accept no, and keep stakeholders informed." During the discussion, the Big Safari Deputy's poker face was firm and unwavering. After I finished presenting my perception of The Big Safari Way an ever-so-slight smile could be seen on my host's face. "Yup, you got it," he said.

A few years later I was assigned to lead the F-22 System Program Office as the Senior Program Director. When the announcement came out identifying a new F-22 Deputy Director, I was excited it was the same Big Safari Deputy Director I had previously met. I knew he would make an immediate positive impact on the team and mission, and I really wanted him to join the leadership team. His passionate opposition to the job was finally overcome when the human resources office told him to take the F-22 Deputy Director job or resign from government service. He took the job.

The two of us made a great leadership team and together we brought significant improvements to the F-22 program. The many accomplishments included rapidly fielding an automatic backup oxygen system, resolving the perceived hypoxia concern, standing up an organic depot, and implementing a long-term sustainment contract. All of this was done, by the way, in the face of "the Pentagon will never approve it" opposition. Nonetheless, the team persevered and eventually navigated their way to a whopping one-billion dollars in cost savings for the American taxpayers. For the first time in the history of the F-22 program the budget requests were declining while project outputs were increasing. Just as importantly, Team F-22 was setting and meeting commitments across the board to recapture stakeholder and warfighter trust and confidence.

After a year of helping the F-22 team solve some of the biggest issues in the history of the F-22 Raptor program, the reluctant deputy decided to retire from civil service. True to form he initially rejected any attempt to organize a retirement ceremony. But my nagging finally irritated him enough and he agreed to a "short" celebratory luncheon.

After the glowing words about his awesome military and civilian careers, his many accomplishments at Big Safari, and his positive effect on team F-22, it was time for him to speak. His words continue to resonate.

"A few years ago, I got a call from my son, who was deployed in Afghanistan fighting for the good guys. He started the call by reminding me of all the times I was away from the family while he was growing up. The birthdays I missed. The long-hours and late nights I was at work while he ate dinner at home without me at the table."

"Just as I was feeling bad about all the things I had missed, and feeling a failure as his dad, he said, 'but I get it now.' Then he told me about a battle he had just survived. It was an unexpected attack, but by good fortune an American MC-12 Liberty reconnaissance aircraft was in the area. Some of you know I was the Liberty program director at Big Safari, and we modified a civilian aircraft with surveillance and reconnaissance gear

and fielded the capability in less than one-hundred-and-eighty days."

"My son told me the Liberty entered the airspace and started calling in mortar and air strikes on the enemy, and the American's wiped the enemy out."

"Dad, my buddies and I are alive today because of that system. We are alive because of the sacrifices you made to get Liberty and systems like it out here. Thank you, Dad, I love you."[15]

Tears welled in his eyes as he finished this incredible story. They were not tears of sorrow, but of pride. Pride in being a small part of an incredibly important profession, a profession that helps our nation win wars and saves lives.

The F-22 Deputy's story is a case-in-point on why Project Management Professionals and their teammates must work tirelessly to rapidly deliver capability. The systems you work on matter, they help our warriors fight with lethality and they save lives. Never forget that.

The go fast, don't accept no, keep leadership informed culture of Big Safari can be replicated in any product development setting. As a program executive explained, there are no special authorities in Big Safari, just special people. People who put the product ahead of the process, find the existing authorities needed to rapidly deliver capability, and do so without excuses. The ten principles in this book are the blueprint on how to lead your project this way.

Become the leader who recognizes the real problems in the bureaucracy and learns to plow down roadblocks or navigate around them. A professional who understands success is all about rapidly delivering the needed capability on-time and on-cost. A trailblazer who demands achievable requirements to leverage existing technology, moves inventions to laboratories, and avoids scope creep. A contrarian who captures trade space by recognizing design stability is much more important than requirements stability. The data-driven technician who keeps stakeholders informed to avoid disruptive, time-consuming, and costly surprises. A passionate coach articulating and proliferating their project's intent thereby empowering the team at all levels to make the decisions required for program success. An entrepreneur with the moral courage to assume leadership when needed to accelerate decision making, and diligently meet commitments to build trust, confidence, and advocacy. Finally, a compassionate and empathetic leader who places delivering for their customers ahead of all else.

These are the ten principles underpinning the mindset of high-performing project managers. The go fast, don't accept no, keep leadership informed project management mavericks and their highly successful product development teams. The tenets can and should be used by every project management professional irrespective of specialty or industry. This is the mindset on how we meet our nation's call as project management professionals to "accelerate change or lose."[16]

111

If you called me up for advice on how to manage a project, I would first tell you to avoid the bureaucratic quicksand by becoming laser-focused on rapidly delivering the needed capability on-time and on-cost. Then I'd cover the ten principles included in this purposefully short book. Finally, I'd tell you this: Thank You!

Thank you for volunteering to serve as a project management professional. Thank you for working tirelessly to deliver war-winning capabilities that give our nation's warriors an unfair technical advantage so they can fight with deadliness, decisively win, and return home safely. Thank you or delivering value-added products to enhance our nation's economic instrument of power. Finally, thank you for being a project management professional. You are making a positive difference for our nation and there are millions of Americans like the Pearson's who appreciate you. Now, get out there and DELIVER for your customers or our nation's warfighters!

> *Out of intense complexities,*
> *intense simplicities emerge.*
> - Sir Winston Churchill

Chapter Endnotes

---

[15] Special thanks to my friend and former Deputy for his permission to share this story.

[16] Quote by General Charles Q. Brown, as Chief of Staff of the United States Air Force. General Brown is currently the advisor to the military advisor to the President of the United States as sitting Chairman of the Joint Chiefs of Staff.

Afterword: Become a Program Leader

Program managers are certified; program leaders are qualified

A subtle theme runs throughout this book. Moving from project manager to program leader requires years of additional self-paced and experience-based learning. This is true of many professions. For example, completing the eight years of post-high school education to graduate from medical school does not make one a doctor. Instead, it earns them the title of resident, or doctor in training. A medical resident is under the constant supervision of a fully qualified doctor for an additional three-years of on-the-job training. Only after successfully completing residency will the medical community acknowledge the doctor in-training has become a qualified doctor. A doctor who is fully authorized to diagnose and treat patients without supervision. It's a grueling decade long process.

Like the medical resident the certified Program Management Professional has a four-year college degree plus three additional years of relevant experience. Seven years to get certified in the profession's vocabulary, policies, and processes. However, completing seven years of post-high school education and experience only produces a program leader in training, a project resident of sorts. Graduating to program leader requires a determined focus to capture the four additional competencies required to become a qualified program leader. Qualified program leaders know how to manage technical programs, lead people, navigate bureaucracy, and unify stakeholders. These are the four essential competencies differentiating certified program managers from qualified program leaders.

Program leaders in training target experiences starting with small projects and incrementally building up to larger projects or portfolios. In doing so they learn about requirements generation and strategy development. They learn how to build integrated schedules, craft metrics, drive accountability, and deliver value-added products. By capturing this additional knowledge through targeted experiences, program leaders in-training begin to learn how to effectively and efficiently *manage technical projects and baselines.*

Program leaders in-training are opportunists who practice leadership wherever and whenever possible. They organize community events, fundraising drives, or even coach youth sports. This is how they practice developing others and motivating a team toward a common end-state. They work on their public speaking skills by volunteering to give lectures and speeches throughout the community. They practice writing to learn how to succinctly and unambiguously craft a meaningful message. Program leaders in training constantly work on self-improvement to mature their strengths and improve upon their weaknesses. These are a few of the ways program leaders in training begin capturing the skills necessary to successfully and empathetically *lead people*.

Program leaders in training seek out assignments at their corporate headquarters or the Pentagon. Here they learn how to capture program resources and identify the types of relevant information executives need to make data-driven decisions. They learn by witnessing real-time case studies of project successes and failures. By gaining an appreciation for the needs of those at the top of an organization, program leaders in training develop a personal methodology for how to effectively *navigate bureaucracy*.

Program leaders in-training practice communicating with, and unifying, stakeholders. They learn to filter out interesting from relevant information. They work to establish regular governance forums and build periodic reports. All of this focused experience to learn how to *unify stakeholders*.

Program leaders in training also conduct a self-paced continuing education program. A program focused on learning as much as possible about managing programs, leading people, navigating bureaucracy, and unifying stakeholders. They study books like *"The Mind of War"* to learn what it takes to properly navigate bureaucracies. They peruse *"Augustine's Laws"* to see how well-written paper promises can win a product development contract, only to turn into cost and schedule nightmares. They study *"The Machine that Changed the World"* and elevate their understanding of how continuous process improvement can drive down costs and expedite schedules.

These types of professionals agree with President Harry Truman, "all leaders are readers." A recommended reading list is located in an appendix to this book.

Finally, program leaders in training reflect on their self-education and targeted experiences to learn, grow, and develop into dependable leaders. They perform the gut checks necessary to move their learning into new and lasting personal behaviors and habits.

Targeted experiences, self-initiated education, and self-reflection are the keys to evolving into a highly effective program leader. A leader who knows how to manage technical programs, lead people, navigate bureaucracy, and unify stakeholders. Evolving from certified project manager to qualified program leader is a marathon, not a sprint.

The ten principles in this book are your roadmap to capturing the mindset of a high performing program leader. Use this book to guide your company, or our country, to victory.

> *A nation's ability to fight a modern war*
> *is as good as its technological ability.*
> - Air Commodore Sir Frank Whittle

Appendix A: Recommended Reading

Augustine, Norm R. *Augustine's Laws*. Penguin Books. New York, NY. 1983.

Burton, James G. *The Pentagon Wars: Reformers Challenge the Old Guard*. Naval Institute Press. Annapolis, MD. 1993.

Charan, Ram, and Geoffrey Colvin. *Fortune Magazine.* "Why CEOs Fail?" June 21,1999.

Drewes, Robert W. *The Air Force and the Great Engine War*. National Defense University Press. Washington D.C. 2005.

Fox, Ronald J., David G Allen, Thomas C. Lassman, Walton S. Moody, and Philip L. Shiman. *Defense Acquisition Reform 1960-2009: An Elusive Goal*. United States Army. Washington D.C., 2011.

Gansler, Jacques S. *Affording Defense*. MIT Press. Cambridge, MA. 1991.

Hammond, Grant T. *The Mind of War: John Boyd and American Security*. Smithsonian Institution Press. Washington D.C., 2001.

Kadish, Ronald T. and Gerald Abbott, Frank Cappuccio, Richard Hawley, Paul Kern, Donald Kozlowsi. *Defense Acquisition Performance Assessment Report.* Office of the Deputy Secretary of Defense. Washington D.C., 2006. Available online at Defense Technical Information Center (apps.dtic.mil).

Mandeles, Dr. Mark D. *The Development of the B-52 and Jet Propulsion: A Case Study in Organizational Innovation.* Air University Press. Maxwell AFB, AL. 1998.

Morgan, James M., and Jeffrey K. Liker. *The Toyota Product Development System. Integrating People, Process, and Technology.* Productivity Press, NY. 2006.

Rasor, Dina, with Sharyl B. Cohen and Donna Martin. *More Bucks Less Bang: How the Pentagon Buys Ineffective Weapons.* Fund for Constitutional Government. Washington D.C., 1983.

Rich, Ben, and Leo Janos. *Skunk Works.* Little Brown and Company, Back Bay Books. New York, NY. 1994.

Younossi, Obaid, and Mark V. Arena, Robert S. Leonard, Charles Robert Roll Jr, Arvind Jain, and Jerry M. Sollinger. *Is Weapon System Cost Growth Increasing? A Quantitative Assessment of Completed and Ongoing Programs.* RAND Project Air Force. 2007.

Appendix B: Program Leader Competencies

gut check (noun): a test or assessment of courage, character, or determination (Merriam-Webster Online Dictionary)

Gutt Check™ (noun): Leverage decades of experience and education to reflectively determine the most effective methods for accomplishing meaningful duties.

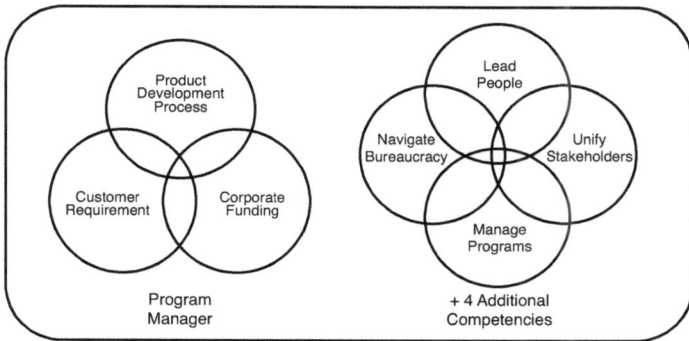

Gutt Check: Program Leader

Program Leader (noun): 1. Project Management Professionals who are both certified and qualified. 2. A certified project manager who has captured, through self-initiated education, experience, and reflection the ability to lead people, navigate bureaucracy, manage projects, and unify stakeholders.

Appendix C: Commitment Tracking Example

Vest Valve Replacement Project As of X December 20XX
Project Manager: John Doe **Customer:** Ms Smith, Company A **Budget/Value:** $2 million **Need Date:** 31 December 2012
Description/Deliverables •Pilot vest pressure valve open; must close during normal flight •Deliver 500 proper functioning valves by 31 Dec 2012 for $2M
Status / Accomplishments •Lab test report complete •Flight Test report on track
Risks If flight test does not conclude on time, then fielding will be delayed

Project Milestones	Baseline	Revised	Actual
Requirements Review	06/19/12		06/19/12
Prototypes Selected	07/06/12		07/03/12
Ground Testing Begins	07/13/12		07/13/12
Ground Testing Ends	08/27/12		08/20/12
Valve selected	09/14/12	09/12/12[2]	09/12/12
Initial test units delivered	09/21/12		09/06/12
Flight Testing Begins	11/06/12	12/03/12[6]	12/03/12
Fight Testing Ends	11/14/12	12/08/12[6]	12/06/12
Flight Test Report Complete	12/15/12	12/20/12[6]	
Decision to Field	09/17/12	12/27/12[7]	
Documentation Updated	11/05/12		12/05/12
Valves Delivered	11/28/12	12/15/12[6]	
Spares Delivered	12/07/12	12/20/12[5]	
Project Complete	12/30/12		

*Superscript = # times date changed since initiation

Description/Deliverables: Identify the problem statement and the Project Management Baseline. The baseline should answer "We plan to _(do what)_ by _(when/date)_ for _(this much budget/money)_."

Inch stones: The right side of the chart is your schedule of tasks and milestones. Tasks should be no more than thirty to sixty days apart to achieve proactive schedule management.

Appendix D: Summary

Deliver!
The Mindset of High-Performing Project Managers

- Learn to navigate: Identify and avoid the bureaucratic roadblocks

- Define success: focus on product ahead of process

- Demand achievable requirements; leverage existing technology; dodge inventing; and avoid scope creep

- Capture trade space: design stability is more important than requirements stability

- Leverage competition: companies want to win a lifetime worth of revenue; avoid paper promises, make them compete for it

- Keep stakeholders informed: turn critics into advocates

- Articulate your intent: empower the team at all levels

- Be decisive: in the absence of leadership, assume it

- Meet commitments: proactively manage schedules, build trust and confidence, remain credible

- Find your "Why": Remain passionate about your higher-level purpose

Bibliography

B.H. Liddell Hart. *Thoughts on War*. London: Farber, 1944. Pages 158, 164. As summarized by: *Air Force Association*. Quotations on Airpower. Reviewed March 2023. https://secure.afa.org/quotes/quotes.pdf.

Contrails. Air Force Academy Cadet Handbook. Volume 31. The United States Air Force Academy, Colorado. 1985-1986. Page 183.

Charan, Ram, and Geoffrey Colvin. *Fortune Magazine*. Why CEOs Fail? June 21,1999.

Drewes, Robert W. *The Air Force and the Great Engine War*. National Defense University Press. Washington D.C. 2005.

Fox, Ronald J, with with David G Allen, Thomas C. Lassman, Walton S. Moody, and Philip L. Shiman. *Defense Acquisition Reform 1960-2009: An Elusive Goal*. Center of Military History. United States Army. Washington D.C., 2011. Page 147.

Gutterman, Greg. *Hustle and Have Fun! A Coach's Guide to Winning Over Players and Parents*. Greg Gutterman Group, Ohio. 2023.

Matis, Jim with Bing West. *Call Sign Chaos: Learning to Lead*. Random House, New York. 2019.

United States Government Accountability Office. Report to Congressional Committees. Weapons Systems Annual Assessment: Challenges to Fielding Capabilities Faster Persist. GAO-22-105230. June 2022. Pages 27-28.

United States Government Accountability Office. Defense Acquisitions Annual Assessment: Drive to Deliver Capabilities Faster Increases Importance of Program Knowledge and Consistent Data for Oversight. GAO-20-439. June 3, 2020. Reviewed March 2023. https://www.gao.gov/products/gao-20-439

Acknowledgments

First and foremost, to Leslie and our children. You are my oxygen and reason for everything.

Mom and Dad. Despite your escape from earthly bonds, the lessons you taught me are clearer every day.

My mentors and advocates. Thank you!

My editor, Larkin, and numerous reviewers. A debt of gratitude I will be unable to repay for all your thoughts, comments, and edits. Thank you!

My Air Force. I am grateful beyond belief to have been a small part of the world's most powerful Air Force.

My military brethren. We have the most capable military the world has ever known, because you are the most incredibly talented and dedicated warriors on the planet. I was honored to serve alongside you!

Our nation's lost warriors and their families; heroes all.

Finally, thank you to Army Corporal Samuel Pearson from Piqua, Ohio. Our gratitude forever.

Books and Contact Information

Read the book Hall of Fame National Basketball Association coach **Gregg Popovich** said, *"If you're a coach or parent, read this book!"* **Hustle and Have Fun! A Coach's Guide to Winning Over Players and Parents** is General-Coach Gutterman's blueprint for teaching character through sports. If you volunteered to coach a youth sports team and are wondering "now what?" Or, as a parent you're asking, "what should I expect from my kids coach?" If your answers are "Yes," then go to Amazon.com for your copy today.

Look for future Gutt Check series books by General Gutterman on the topics of leadership, life, and more.

For information, workshops, speaking engagements, and more contact General Gutterman at CoachGutt@HustleandHave.Fun.

About the Author

United States Air Force Brigadier General Greg Gutterman, retired, is a native of Minnesota. He graduated from the United States Air Force Academy and spent three decades as an active-duty officer.

The general served as an engineer and program leader on every product group and at nearly every echelon in the United States Air Force's technical hierarchy. His experience spans the F-22 Raptor stealth fighter, enterprise software, intelligence systems, international sales, and more.

General Gutterman is now preparing the next generation of acquisition warriors as an instructor at the Air Force Institute of Technology in Dayton, Ohio.